T0212234

Analysis and Design of Substrate Integrated Waveguide Using Efficient 2D Hybrid Method

Synthesis Lectures on Computational Electromagnetics

Editor
Constantine A. Balanis, *Arizona State University*

Synthesis Lectures on Computational Electromagnetics will publish 50- to 100-page publications on topics that include advanced and state-of-the-art methods for modeling complex and practical electromagnetic boundary value problems. Each lecture develops, in a unified manner, the method based on Maxwell's equations along with the boundary conditions and other auxiliary relations, extends underlying concepts needed for sequential material, and progresses to more advanced techniques and modeling. Computer software, when appropriate and available, is included for computation, visualization and design. The authors selected to write the lectures are leading experts on the subject that have extensive background in the theory, numerical techniques, modeling, computations and software development.

The series is designed to:

- Develop computational methods to solve complex and practical electromagnetic boundary-value problems of the 21st century.

- Meet the demands of a new era in information delivery for engineers, scientists, technologists and engineering managers in the fields of wireless communication, radiation, propagation, communication, navigation, radar, RF systems, remote sensing, and biotechnology who require a better understanding and application of the analytical, numerical and computational methods for electromagnetics.

Analysis and Design of Substrate Integrated Waveguide Using Efficient 2D Hybrid Method
Xuan Hui Wu and Ahmed A. Kishk
2010

An Introduction to the Locally-Corrected Nyström Method
Andrew F. Peterson and Malcolm M. Bibby
2009

Transient Signals on Transmission Lines: An Introduction to Non-Ideal Effects and Signal Integrity Issues in Electrical Systems
Andrew F. Peterson and Gregory D. Durgin

2008

Reduction of a Ship's Magnetic Field Signatures
John J. Holmes
2008

Integral Equation Methods for Electromagnetic and Elastic Waves
Weng Cho Chew, Mei Song Tong, and Bin Hu
2008

Modern EMC Analysis Techniques Volume II: Models and Applications
Nikolaos V. Kantartzis and Theodoros D. Tsiboukis
2008

Modern EMC Analysis Techniques Volume I: Time-Domain Computational Schemes
Nikolaos V. Kantartzis and Theodoros D. Tsiboukis
2008

Particle Swarm Optimization: A Physics-Based Approach
Said M. Mikki and Ahmed A. Kishk
2008

Three-Dimensional Integration and Modeling: A Revolution in RF and Wireless Packaging
Jong-Hoon Lee and Manos M. Tentzeris
2007

Electromagnetic Scattering Using the Iterative Multiregion Technique
Mohamed H. Al Sharkawy, Veysel Demir, and Atef Z. Elsherbeni
2007

Electromagnetics and Antenna Optimization Using Taguchi's Method
Wei-Chung Weng, Fan Yang, and Atef Elsherbeni
2007

Fundamentals of Electromagnetics 1: Internal Behavior of Lumped Elements
David Voltmer
2007

Fundamentals of Electromagnetics 2: Quasistatics and Waves
David Voltmer
2007

Modeling a Ship's Ferromagnetic Signatures
John J. Holmes
2007

Analysis and Design of Substrate Integrated Waveguide Using Efficient 2D Hybrid Method

Xuan Hui Wu and Ahmed A. Kishk

ISBN: 978-3-031-00583-1   paperback
ISBN: 978-3-031-01711-7   ebook

DOI 10.1007/978-3-031-01711-7

A Publication in the Springer series
SYNTHESIS LECTURES ON COMPUTATIONAL ELECTROMAGNETICS

Lecture #26
Series Editor: Constantine A. Balanis, *Arizona State University*
Series ISSN
Synthesis Lectures on Computational Electromagnetics
Print 1932-1252   Electronic 1932-1716

Analysis and Design of Substrate Integrated Waveguide Using Efficient 2D Hybrid Method

Xuan Hui Wu
Radio Waves, Inc.

Ahmed A. Kishk
University of Mississippi

SYNTHESIS LECTURES ON COMPUTATIONAL ELECTROMAGNETICS #26

ABSTRACT

Substrate integrated waveguide (SIW) is a new type of transmission line. It implements a waveguide on a piece of printed circuit board by emulating the side walls of the waveguide using two rows of metal posts. It inherits the merits both from the microstrip for compact size and easy integration, and from the waveguide for low radiation loss, and thus opens another door to design efficient microwave circuits and antennas at a low cost. This book presents a two-dimensional fullwave analysis method to investigate an SIW circuit composed of metal and dielectric posts. It combines the cylindrical eigenfunction expansion and the method of moments to avoid geometrical descritization of the posts. The method is presented step-by-step, with all the necessary formulations provided for a practitioner who wants to implement this method by himself. This book covers the SIW circuit printed on either homogeneous or inhomogeneous substrate, the microstrip-to-SIW transition and the speed-up technique for the simulation of symmetrical SIW circuits. Different types of SIW circuits are shown and simulated using the proposed method. In addition, several slot antennas and horn antennas fabricated using the SIW technology are also given.

KEYWORDS

substrate integrated waveguide (SIW), method of moments, cylindrical eigenfunction expansion, microwave, millimeter wave, filter, power divider, ring hybrid, directional coupler, antenna

Contents

CHAPTER 1

Introduction

In microwave engineering, microstrip is widely used in the design of all kinds of passive circuits due to its compact size, easy integration and readiness for mass production. But with frequency increasing, as an open structure, a microstrip circuit shows undesirable radiation. Such radiation not only introduces additional losses in the circuit but it also has a negative impact on the surrounding components. For example, if the feeding network of an antenna array is implemented using a microstrip-type of circuit, the undesired radiation from the feeding network may seriously affect the radiation performance of the antenna. Alternatively, the traditional waveguide circuit has the minimum radiation loss as it is a closed structure and all the electromagnetic energy is bounded inside the waveguide, but it is cumbersome compared to its microstrip counterpart. With frequency increasing, the physical dimensions of the waveguide decreases, but the integration of many waveguide circuits is still not as easy as that for the microstrip circuits.

Recently, a new type of transmission line called substrate integrated waveguide or post-wall waveguide is invented. It is a low cost realization of the traditional waveguide circuit for microwave and millimeter-wave applications. It inherits the merits from both the traditional microstrip for easy integration and the waveguide for low radiation loss. In such a circuit, metallic posts are embedded into a printed circuit board, which is covered with conducting sheets on both sides, to emulate the vertical walls of a traditional waveguide. It has the advantages of the traditional waveguide circuit, such as low radiation loss, high Q-factor and high power capacity while can be readily fabricated with the existing technologies like printed circuit board and low-temperature co-fired ceramic (LTCC) technologies. In addition, integration of many substrate integrated waveguide circuits into a single-board sub-system is also possible. For the metal posts that emulate a vertical wall, the post radius and separation should be carefully designed in order to reduce the wave leakage and to get a desired propagation constant. Specially, for two rows of periodic metal posts, the wave propagation characteristics have been studied, and the design equations for the post walls can be found [1, 2, 3]. However, to study a general substrate integrated waveguide circuit, a fullwave analysis is required.

For most single layer substrate integrated waveguide circuits, the top and bottom copper sheets are intact without slots and they are excited by a TE-like mode electromagnetic waves. Assuming no field variation normal to the substrate, this type of circuits can be considered as a 2D problem with only vertical electric and horizontal magnetic fields if the circuit is placed in the horizontal plane. Finite element method and finite-difference time-domain method can be used to solve it, but they require the geometry discretization of the entire circuit which may require large memory and be time consuming. Method of moments (MOM) that only discretizes the geometry discontinuities may also be used [4]. But, thin wire approximation that assumes a uniform circumferential current density may be invalid for a metal post on the post wall even if it is electrically thin because the current density on the post wall looking from inside of the circuit is much stronger than that looking from outside of the circuit. Thus, the discretization of the metal posts is unavoidable and that increases the system matrix size. Also, in MOM, the boundary condition is enforced at discrete positions on a metal post if point matching is adopted or at the entire surface in an average sense if a testing procedure is applied. Recently, cylindrical eigenfunction was applied to study crystal devices [5, 6, 7, 8] with circular cylindrical elements. An efficient hybrid method is presented [9, 10] to study a 2D substrate integrated waveguide circuit. The field due to a cylinder is written in a series of cylindrical eigenfunctions assuming no variation of the fields along the cylinders, which makes the problem of 2D type, while the waveguide ports are treated in an MOM manner. There is no geometry discretization for the cylinders, and the boundary conditions at the entire surface of a cylinder are enforced intrinsically. Recently, many substrate integrated waveguide circuits have been invented, for example, a six-port junction [11], a millimeter-wave diplexer [12], a filter [13] and a ring hybrid [14]. The substrate integrated waveguide can also be used in an active circuit with the help of an SIW-to-microstrip transition [15].

The substrate integrated waveguide technique can also be used to design antennas. Radiation of the electromagnetic energy can be achieved by cutting slots on the copper sheets over the substrate, or leaving the end of a substrate integrated waveguide open but not closed by a post wall. For these structures, the electromagnetic field is no more constant in the normal direction of the substrate and a 3D electromagnetic solver must be applied. Several novel substrate integrated waveguide antennas were proposed recently. For example, an omnidirectional antenna is made by cutting slots on both sides of a substrate integrated waveguide [16], and an H-plane sectoral horn antenna is formed by flaring the two post walls of a substrate integrated waveguide and leaving the end open [17]. Thanks to the planar fabrication techniques, many radiation elements can be fabricated on a single circuit board to make an antenna array [17, 18]. Such an antenna array requires a feeding network to achieve a desired beam direction. Although a 2D method presented in [9, 10] is not for a radiating structure, it does help to efficiently design the feeding network for a substrate integrated waveguide antenna array.

In this book, the hybrid method is presented in details. The book is organized as follows. Chapter 2 presents a full-wave analysis algorithm to analyze a substrate integrated waveguide circuit composed of metal posts. It combines the method of moments and cylindrical eigenfunction expansion in order to capture the variation of the electromagnetic field around a circular cylinder while at the same time incorporate the contribution due to the non-circular part of the circuit, such as a waveguide port. Several circuit examples are designed using this method and they are verified by a commercial software package [19]. Chapter 3 presents the extended version of the hybrid method in order to deal with a circuit fabricated on an inhomogeneous substrate. Different dilectric cylinders can be inserted into the substrate and thus provide more freedom in the circuit design. Two filter designs are demonstrated using this inhomogeneous substrate technique. Chapter 4 talks about how to speed up the simulation by making use of circuit symmetry. For a symmetrical circuit, only half of the circuit is needed to be analyzed twice instead of simulating the entire structure. PEC or PMC wall is used to close the half circuit. This may require less computation resources since the simulation of two small linear systems may be much faster than the simulation of a large system and need less memory. A symmetrical ring hybrid and a T junction are designed using this speed-up technique. Chapter 5 discusses the transition from a substrate integrated waveguide to a traditional microstrip and the half mode substrate integrated waveguide circuits. The method to study a tapered line transition is presented where the transition is modeled by a PMC wall that follows the boundary of the transition. The half mode substrate integrated waveguide is a compact counterpart of its full structure. It is shown that it can be simulated using the same method for the tapered line transition. Chapter 6 presents two types of antennas based on the substrate integrated waveguide technique. One is the slot antenna and the other one is the H-plane sectoral antenna. Due to the maturity of the planar fabrication technique, an array of either the slot antennas or sectoral horn antennas can be fabricated together with the feeding network on a single printed circuit board. As mentioned before, although the simulation method presented in this book is not for radiating structures, it is still a good tool to design the feeding network for a substrate integrated waveguide antenna array.

CHAPTER 2

SIW Circuits Composed of Metallic Posts

A substrate integrated waveguide circuit with metallic posts only is the simplest and most popular configuration because it can be easily fabricated using the existing printed circuit board technique. The analysis method presented in this chapter focuses on this type of circuits, and it follows a classical procedure of solving an electromagnetic problem. First, the electromagnetic field expressions in the substrate are obtained. Then, the boundary conditions are enforced at the ports and the boundary of each metallic post. After that, a linear system that consists of a set of linear equations is derived from the boundary conditions. Finally, the circuit characteristics, such as the S parameters, can be obtained by solving the linear system. At the end of this chapter, several substrate integrated waveguide circuits are analyzed.

2.1 A TYPICAL SIW CIRCUIT AND ITS EQUIVALENT PROBLEM

Fig. 2.1 shows an x-y cut of a typical two-port substrate integrated waveguide circuit composed of metallic posts with different sizes. The metallic posts A, B and C are located inside the circuit boundary, and those without labels emulate two metallic walls. Without losing generality, the circuit in Fig. 2.1 is used as an example to present the method.

In order to solve the circuit, an approximated equivalent problem is constructed in two steps. First, a closed contour is generated to enclose the circuit as shown in Fig. 2.2(a), where part of the contour coincides with the waveguide ports. Based on the surface equivalent principle, the electromagnetic fields inside the contour are due to the surface equivalent currents on the contour, with the presence of all the cylinders. On the contour, the electric current density $\vec{J}_s = \hat{n} \times \vec{H}$ and the magnetic current density $\vec{M}_s = -\hat{n} \times \vec{E}$, where $\vec{H}$ is the total magnetic field, $\vec{E}$ is the total electric field and $\hat{n}$ is a unit vector normal to the contour and points inside the circuit. Since the post walls are designed to emulate the vertical walls of the traditional waveguide circuit using design rules in [1, 3], the field leakage from the post walls are very weak, and thus the equivalent currents outside the post walls can be ignored. As a special case, if the radius and separation of the metallic posts on post walls are infinitely small, the asymptotic boundary conditions in [20] can be applied. It claims that for period PEC strips with infinitely small period and width, the tangential electrical field along the strips equals to zero. So that, $E^z = 0$ at the post wall and the post wall acts as a solid metallic wall for the assumed propagated mode where $E^x = E^y = 0$. Consequently, the field inside the circuit is due to the equivalent currents at the ports, with the presence of all the cylinders as shown in Fig. 2.2(b).

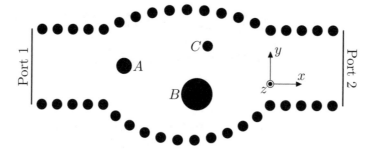

Figure 2.1: A typical SIW circuit composed of metallic posts of different radii.

2.2   FIELD EXPRESSIONS

The expression of electric field is required in order to apply the boundary conditions at all the material discontinuities. For the problem in Fig. 2.2(b), only the region outside the metallic posts have non-zero field, and the electric field can be expressed as

$$E_S^z = E_{\text{port}}^z + \sum_{n=1}^{N} \sum_{m=-M_n}^{M_n} a_{m,n} H_m^{(2)}(\beta_s \rho_n) \exp(jm\phi_n)$$

$$+ \sum_{\kappa=\{A,B,C\}} \sum_{m=-M_\kappa}^{M_\kappa} a_{m,\kappa} H_m^{(2)}(\beta_s \rho_\kappa) \exp(jm\phi_\kappa) \tag{2.1}$$

where $H_m^{(2)}(\cdot)$ is the mth order Hankel function of the second kind, E_{port}^z is the field due to the equivalent currents on the ports, N is the number of the metal posts that comprise the post walls, ρ_n and ϕ_n are the local coordinates of the nth metal post for the post walls, ρ_κ and ϕ_κ are those of the cylinder κ where κ represents A, B or C, and β_s is the phase constant in the substrate. Note that the expansion order m is truncated from $-M_n$ to M_n for the nth metal post comprising the post wall and from $-M_\kappa$ to M_κ for the cylinder κ. The local coordinate system of a cylinder is defined such that the origin is put at the center of the cylinder. For example, Fig. 2.3 illustrates the local coordinate system of the cylinder A.

E_{port}^z in (2.1) can be obtained using MOM as follows. Each waveguide port is divided into several segments, each referred to as a sub-port, and the $\vec{J}_s$ and $\vec{M}_s$ are assumed constant for each sub-port. In a coordinate system as shown in Fig. 2.4, the electric field at an observation position (x, y) and due to the $\vec{J}_s$ on a sub-port can be obtained as

$$E_{J_s}^z(x, y) = \frac{\omega \mu_s I_J}{-4} \int_0^L H_0^{(2)}(\beta_s \rho') dx' \tag{2.2}$$

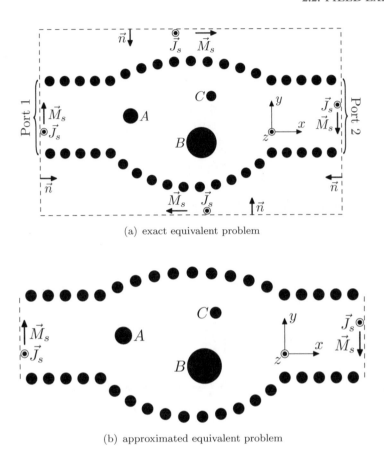

(a) exact equivalent problem

(b) approximated equivalent problem

Figure 2.2: Equivalent problems of a substrate integrated waveguide circuit with metallic posts.

where I_J is the unknown electric current density for the sub-port, $\rho' = \sqrt{(x - x')^2 + y^2}$, L is the sub-port length and μ_s is the permeability of the substrate. Similarly, the electric field at (x, y) and due to $\vec{M}_s$ on a sub-port can be obtained as

$$E^z_{M_s}(x, y) = \frac{j\beta_s I_M}{4} \int_0^L \sin \phi' H_1^{(2)}(\beta_s \rho') dx' \qquad (2.3)$$

where I_M is the unknown magnetic current density, $\sin \phi' = y/\rho'$ and ρ' is the same as that for (2.2). With (2.2) and (2.3), E^z_{port} in (2.1) can be obtained by adding the contributions of all the sub-ports. If (x, y) in Fig. 2.4 is not located on the sub-port itself, (2.2) and (2.3) can be evaluated using numerical integration. Otherwise, they should be evaluated in a different way as follows. Suppose the observation point in Fig. 2.4(a) is located on the sub-port itself, at $(x, 0)$ with $x \in (0, L)$, (2.2)

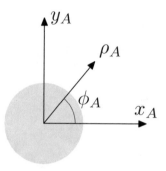

Figure 2.3: Local coordinate system of the cylinder A.

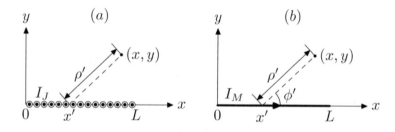

Figure 2.4: Radiation of a sub-port due to (a) J_s and (b) M_s; from [10], copyright ©2008 IEEE.

can be written as

$$
E_{J_s}^z(x, 0) = \frac{\omega \mu_s I_J}{-4} \left\{ \int_0^L \left(1 - \frac{2j}{\pi} \ln \frac{\gamma \beta_s \rho'}{2} \right) dx' \right.
$$
$$
\left. + \int_0^L \left[H_0^{(2)}(\beta_s \rho') - 1 + \frac{2j}{\pi} \ln \frac{\gamma \beta_s \rho'}{2} \right] dx' \right\} \tag{2.4}
$$

where $\gamma = 1.781$. The second integration can be computed numerically as its integrand is bounded while the first one should be evaluated analytically as

$$
\int_0^L \left(1 - \frac{2j}{\pi} \ln \frac{\gamma \beta_s \rho'}{2} \right) dx'
$$
$$
= L - \frac{2j}{\pi} \left[\int_0^x \ln \frac{\gamma \beta_s (x - x')}{2} dx' + \int_x^L \ln \frac{\gamma \beta_s (x' - x)}{2} dx' \right]
$$
$$
= L - \frac{2j}{\pi} \left\{ x \left(\ln \frac{x \beta_s \gamma}{2} - 1 \right) + (L - x) \left[\ln \frac{(L - x)\beta_s \gamma}{2} - 1 \right] \right\}. \tag{2.5}
$$

The electric field due to a non-zero magnetic current $\vec{M}_s$ in Fig. 2.4(b) is discontinuous across the sub-port. For an observation point very close to the sub-port, (2.3) can be simplified as

$$E_{M_s}^z(x, 0^+) = -I_M/2$$
$$E_{M_s}^z(x, 0^-) = I_M/2 \tag{2.6}$$

where $x \in (0, L)$. In (2.6), 0^+ and 0^- are positive and negative, respectively, numbers that both are infinitely close to zero. Intuitively, $(x, 0^+)$ with $x \in (0, L)$ defines an observation point exactly on the left side of the magnetic current, and $(x, 0^-)$ with $x \in (0, L)$ defines an observation point on the right side of the magnetic current. Equations (2.6) can be intuitively derived as follows. The boundary condition of the Maxwell equations at the magnetic current in Fig. 2.4(b) is

$$-\hat{y} \times \left[E^z(x, 0^+) - E^z(x, 0^-) \right] = \hat{x} I_M. \tag{2.7}$$

Because $E^z(x, 0^+)$ and $E^z(x, 0^-)$ are evaluated symmetrically at the opposite sides of the magnetic current, they must be either equal or out of phase to each other. If they are equal to each other, I_M in (2.7) equals zero, and it is contradict to the assumption of non-zero current. Therefore, $E^z(x, 0^+) = -E^z(x, 0^-)$ which results in (2.6).

2.3   BOUNDARY CONDITIONS

Boundary conditions on the material discontinuities are applied in order to construct a linear system, from which the circuit characteristics can be extracted. For the problem in Fig. 2.2(b), two different types of boundary conditions are enforced. First, the tangential electric and magnetic fields at the boundary of a metal post are related using the surface impedance of the metal post as

$$\frac{E_S^z}{H_S^{\phi_A}} = R_A \text{ at } \rho_A = r_A$$
$$\frac{E_S^z}{H_S^{\phi_B}} = R_B \text{ at } \rho_B = r_B$$
$$\frac{E_S^z}{H_S^{\phi_C}} = R_C \text{ at } \rho_C = r_C$$
$$\frac{E_S^z}{H_S^{\phi_n}} = R_n \text{ at } \rho_n = r_n \text{ with } n = 1, \cdots, N \tag{2.8}$$

where $H_S^{\phi_\kappa}$ with κ representing A, B or C is the ϕ component of the magnetic field in the local coordinate system of the cylinder κ. r_κ and R_κ are the radius and the surface impedance of the cylinder κ, respectively. $H_S^{\phi_n}$, r_n and R_n have the similar notations but are for the nth metal post on the post walls. Specifically, for a perfectly conducting metallic post, the surface impedance equals to zero. The second type of boundary conditions are enforced at the ports. At the center of each sub-port, the $\hat{z}$ directed electric field equals to the equivalent magnetic current density as

$$E_S^z = I_M \tag{2.9}$$

if I_M is defined in the clockwise direction. Note that there is no boundary conditions enforced on the enclosing box other than the ports.

Equation (2.9) is ready to write into a linear combination of the unknown coefficients by substituting (2.1) into (2.9), but the equations in (2.8) are not. Although the electric field expressions required for (2.8) are given in (2.1), equation (2.1) are written as a combination of terms in different local coordinate systems and can not be used directly. Instead, they should be rewritten in the local coordinate system of the cylinder where the boundary conditions are applied. This can be done using the additional theorem of Bessel and Hankel functions. Fig. 2.5 shows the relative location of a gray

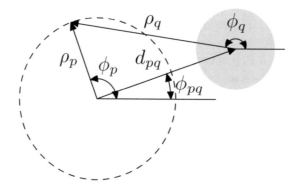

Figure 2.5: Transformation of the local coordinate system from the source cylinder to the observation cylinder.

cylinder and a cylinder in dashed line. The gray cylinder, referred to source cylinder q, scatters field, while the cylinder in dashed line, referred to observation cylinder p, is the one where the boundary conditions are applied. For the relative location of the source and observation cylinders as shown in Fig. 2.5, the electric field due to the source cylinder can be transformed into the local coordinate system of the observation cylinder using

$$H_n^{(2)}(\beta_s \rho_q) \exp(jn\phi_q) = \sum_{m=-Mp}^{Mp} J_m(\beta_s \rho_p) H_{m-n}^{(2)}(\beta_s d_{pq}) \exp(jm\phi_p) \Psi \tag{2.10}$$

where $\Psi = \exp[j(n-m)\phi_{pq}]$. d_{pq} and ϕ_{pq} are the magnitude and phase of a vector pointing from the center of the observation cylinder p to that of the source cylinder p. ρ_p and ϕ_p are the local coordinates of the observation cylinder p, and ρ_q and ϕ_q are those of the source cylinder q. With (2.10), equation (2.1) can be rewritten in the local coordinate system of the observation cylinder p as

$$E_S^z(\rho_p, \phi_p) = \sum_{m=-M_p}^{M_p} U_m^p(\rho_p) \exp(jm\phi_p) \tag{2.11}$$

where $U_m^p(\rho_p)$ is a linear combination of the unknown coefficients and are independent of ϕ_p. Note that the super script p is not an exponent, but an indication that the expression is dependent on the choice of the observation cylinder p. The transformation of (2.2) and (2.3) required for (2.1) follows

$$E_{J_s}^z(\rho_p, \phi_p) = \frac{-\omega\mu_s I_J}{4} \sum_{m=-M_p}^{M_p} \exp(jm\phi_p) \int_0^L J_m(\beta_s\rho_p) H_m^{(2)}(\beta_s d_{px'}) \exp(-jm\phi_{px'}) dx' \tag{2.12}$$

$$
\begin{aligned}
&E_{M_s}^z(\rho_p, \phi_p) \\
&= \frac{\beta_s I_M}{-8} \int_0^L [\exp(j\phi') - \exp(-j\phi')] H_1^{(2)}(\beta_s\rho') dx' \\
&= \frac{\beta_s I_M}{-8} \int_0^L \left[\sum_{n=\{-1,1\}} \exp(jn\phi') H_n^{(2)}(\beta_s\rho') \right] dx' \\
&= \frac{\beta_s I_M}{-8} \sum_{m=-M_p}^{M_p} \exp(jm\phi_p) \int_0^L J_m(\beta_s\rho_p) \left[\sum_{n=\{-1,1\}} H_{m-n}^{(2)}(\beta_s d_{px'}) \exp[j(n-m)\phi_{px'}] \right] dx'
\end{aligned}
\tag{2.13}
$$

where $d_{px'}$ and $\phi_{px'}$ are the magnitude and phase of a vector pointing from the center of the cylinder p to $(x', 0)$ in Fig. 2.4. The integrations in (2.12) and (2.13) can be computed numerically. After the electric field is transformed, the ϕ component of the magnetic field in the local coordinate system of the cylinder p can be obtained using

$$H_S^{\phi_p}(\rho_p, \phi_p) = \frac{-j}{\omega\mu} \frac{\partial E_S^z(\rho_p, \phi_p)}{\partial \rho_p} = \frac{-j}{\omega\mu} \sum_{m=-M_p}^{M_p} \exp(jm\phi_p) \frac{\partial U_m^p(\rho_p)}{\partial \rho_p} \tag{2.14}$$

and is also a series in terms of $\exp(jm\phi_p)$. Consequently, with the orthogonality of $\exp(jm\phi_p)$ for different m, a number of $2M_p + 1$ linear equations can be obtained from each equation in (2.8). For example, the first boundary condition in (2.8) gives $2M_A + 1$ linear equations as

$$U_m^A(r_A) = \frac{-jR_A}{\omega\mu} \frac{\partial U_m^A(\rho_A)}{\partial \rho_A}\bigg|_{\rho_A=r_A} \tag{2.15}$$

where $-M_A \le m \le M_A$. The other boundary conditions in (2.8) can be treated in a similar manner.

2.4 Z MATRIX

For each metallic cylinder, the number of equations obtained by enforcing the boundary conditions at that cylinder is equal to the number of unknowns associated with the same cylinder. However, for each sub-port, there are two unknowns of I_J and I_M but only one equation (2.9). Subsequently, the linear equations can be written in a matrix form as

$$\begin{bmatrix} A_{n\times 2n} & B_{n\times m} \\ C_{m\times 2n} & D_{m\times m} \end{bmatrix} \begin{bmatrix} J_{n\times 1} \\ M_{n\times 1} \\ X_{m\times 1} \end{bmatrix} = 0_{(n+m)\times 1} \tag{2.16}$$

where n is the number of sub-ports, $J_{n\times 1}$ is a vector of the electric current densities on the sub-ports, $M_{n\times 1}$ is a vector of the magnetic current densities on the sub-ports, $X_{m\times 1}$ is a vector of all the other unknowns. Using matrix partition, the above system can be separated into two groups as

$$A_{n\times 2n} \begin{bmatrix} J_{n\times 1} \\ M_{n\times 1} \end{bmatrix} + B_{n\times m} X_{m\times 1} = 0_{n\times 1} \tag{2.17}$$

$$C_{m\times 2n} \begin{bmatrix} J_{n\times 1} \\ M_{n\times 1} \end{bmatrix} + D_{m\times m} X_{m\times 1} = 0_{m\times 1}. \tag{2.18}$$

Solving for $X_{m\times 1}$ from (2.18) and substituting the result into (2.17) gives

$$(A - BD^{-1}C)_{n\times 2n} \begin{bmatrix} J_{n\times 1} \\ M_{n\times 1} \end{bmatrix} = 0_{n\times 1}. \tag{2.19}$$

With matrix partition as $(A - BD^{-1}C)_{n\times 2n} \stackrel{\triangle}{=} [P_{n\times n} \ \ Q_{n\times n}]$, (2.19) leads to

$$M_{n\times 1} = -Q_{n\times n}^{-1} P_{n\times n} J_{n\times 1}. \tag{2.20}$$

Since $M_{n\times 1}$ and $J_{n\times 1}$ are proportional to the voltages and currents, respectively, at the sub-ports, (2.20) can be rewritten in terms of voltages and currents as

$$\frac{-V_{n\times 1}}{h} = -Q_{n\times n}^{-1} P_{n\times n} \mathrm{diag}\left(\frac{1}{l_1}, \cdots, \frac{1}{l_n}\right) I_{n\times 1} \tag{2.21}$$

where $\mathrm{diag}(\cdots)$ is a diagonal matrix, $V_{n\times 1}$ is the sub-port voltages, $I_{n\times 1}$ is the sub-port currents, h is the circuit thickness and l_t with $t = 1, 2, \cdots, n$ is the width of the tth sub-port. Now, the Z matrix is

$$Z_{n\times n} = h Q_{n\times n}^{-1} P_{n\times n} \mathrm{diag}\left(\frac{1}{l_1}, \cdots, \frac{1}{l_n}\right). \tag{2.22}$$

2.5   *S* MATRIX

With the Z matrix obtained before, the S matrix can be easily obtained providing the impedance of all the sub-ports as follows. For a network with n sub-ports as shown in Fig. 2.6, the forward scattering parameter a_n and backward scattering parameter b_n are defined as

$$\begin{cases} a_n = \dfrac{V_n^+}{\sqrt{R_n}} = I_n^+ \sqrt{R_n} \\ b_n = \dfrac{V_n^-}{\sqrt{R_n}} = I_n^- \sqrt{R_n} \end{cases} \tag{2.23}$$

where V_n^+ and I_n^+ are forward voltage and current at the nth sub-port, respectively, and V_n^- and I_n^- are those for the backward wave. The voltage and current at the sub-ports are related using the Z

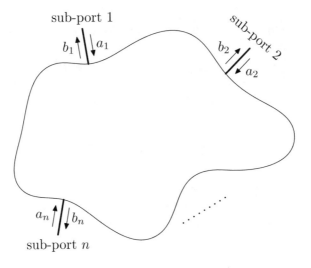

Figure 2.6: A network with n sub-ports.

matrix as

$$V_{n\times1} = Z_{n\times n} I_{n\times1}. \tag{2.24}$$

The voltage and current can be expressed in terms scattering parameters as

$$V_{n\times1} = \text{diag}\left(\sqrt{R_1}, \cdots, \sqrt{R_n}\right)(a_{n\times1} + b_{n\times1}) \tag{2.25}$$

$$I_{n\times1} = \text{diag}\left(\frac{1}{\sqrt{R_1}}, \cdots, \frac{1}{\sqrt{R_n}}\right)(a_{n\times1} - b_{n\times1}) \tag{2.26}$$

where R_n is the impedance for the nth sub-port. Substitute (2.25) and (2.26) into (2.24) gives

$$\text{diag}(\sqrt{R_1}, \cdots, \sqrt{R_n})(a_{n \times 1} + b_{n \times 1})$$
$$= Z_{n \times n} \text{diag}\left(\frac{1}{\sqrt{R_1}}, \cdots, \frac{1}{\sqrt{R_n}}\right)(a_{n \times 1} - b_{n \times 1}). \tag{2.27}$$

Subsequently, $b_{n \times 1}$ can be written in terms of $a_{n \times 1}$ by rearranging (2.27) as

$$b_{n \times 1} = \left[\text{diag}(\sqrt{R_1}, \cdots, \sqrt{R_n}) + Z_{n \times n} \text{diag}\left(\frac{1}{\sqrt{R_1}}, \cdots, \frac{1}{\sqrt{R_n}}\right)\right]^{-1}$$
$$\cdot \left[Z_{n \times n} \text{diag}\left(\frac{1}{\sqrt{R_1}}, \cdots, \frac{1}{\sqrt{R_n}}\right) - \text{diag}(\sqrt{R_1}, \cdots, \sqrt{R_n})\right] a_{n \times 1}. \tag{2.28}$$

Therefore, the S matrix can be computed as

$$S_{n \times n} = \left[\text{diag}(\sqrt{R_1}, \cdots, \sqrt{R_n}) + Z_{n \times n} \text{diag}\left(\frac{1}{\sqrt{R_1}}, \cdots, \frac{1}{\sqrt{R_n}}\right)\right]^{-1}$$
$$\cdot \left[Z_{n \times n} \text{diag}\left(\frac{1}{\sqrt{R_1}}, \cdots, \frac{1}{\sqrt{R_n}}\right) - \text{diag}(\sqrt{R_1}, \cdots, \sqrt{R_n})\right]. \tag{2.29}$$

The impedance of the nth sub-port depends on the excited mode in the waveguide port and can be obtained as

$$R_n = \frac{h}{l_n} Z_{\text{TE}} = \frac{hk\eta}{l_n \beta} \tag{2.30}$$

where Z_{TE} is the wave impedance of the rectangular waveguide port for the excited mode, l_n is the width of the sub-port, h is the thickness of the substrate, k is the wave number in the substrate, η is the characteristic impedance of the substrate and β is the propagation constant in the waveguide port.

2.6   SUB-PORTS COMBINATION

The S matrix obtained before is for the network with all the sub-ports as its ports. Those sub-ports should be combined into waveguide ports to get the final S matrix. As an example shown in Fig. 2.7, two waveguide ports A and B are composed of sub-ports from 1 to m and $m + 1$ to n, respectively. Port A has a width of d_A and Port B has a width of d_B. Both the ports are extended along the x axis and start from $x = 0$. The magnitude of S_{BA} can be obtained by calculating the power transmission from Port A to Port B as

$$|S_{BA}| = \sqrt{\sum_{j=m+1}^{n} \left|\sum_{i=1}^{m} \sqrt{p_i} S_{ji}\right|^2} \quad \text{with} \quad \sum_{i=1}^{m} p_i = 1 \tag{2.31}$$

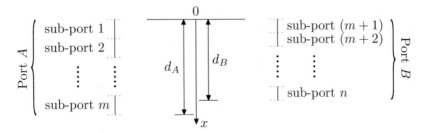

Figure 2.7: Sub-ports combination; from [10], copyright ©2008 IEEE.

where S_{ji} is the jth row and ith column of the S matrix obtained before. p_i in (2.31) is the ratio of the power exciting the ith sub-port to the total power exciting Port A and it is determined by the mode excited in Port A. Specifically, if TE_{10} mode is excited,

$$p_i = \frac{l_i \sin^2(x_i \pi / d_A)}{\sum\limits_{j=1}^{m} l_j \sin^2(x_j \pi / d_A)} \tag{2.32}$$

for $i = 1, 2, \cdots, m$ where l_i and x_i are the width and the center coordinate of the ith sub-port. Assuming Port B is far enough away from the circuit discontinuity, the phase of S_{BA} is equal to that of the transmission coefficient from Port A to any sub-port on Port B and can be obtained as

$$\arg(S_{BA}) = \arg\left(\sum_{i=1}^{m} \sqrt{p_i} S_{ji}\right) \tag{2.33}$$

where j is the index of an arbitrary sub-port on Port B. Practically, for TE_{10} mode, j can be chosen as the index of the sub-port at the center of Port B for high numerical accuracy. Similarly, the reflection coefficient S_{AA} can also be obtained. $|S_{AA}|$ can be obtained in (2.31) by replacing $m + 1$ and n by 1 and m, respectively. The phase of S_{AA} can be obtained in (2.33) by choosing j to be the index of a sub-port on Port A.

2.7   EXAMPLES

The method is implemented to analyze a general substrate integrated waveguide circuit. The truncation of the expansion order for a cylinder κ is chosen as

$$M_\kappa = \text{Int}(3\beta a_\kappa + 1.5) \tag{2.34}$$

where the operator $\text{Int}(\cdot)$ gives the integer part of a real number, a_κ is the radius of the cylinder and β is the real part of the phase constant in the substrate. Several circuits are investigated using this method, and the results are verified by simulating each circuit with its physical thickness using HFSS. In the examples below, TE_{10} mode is assumed for all the ports, and zero surface impedance is assumed for each metallic post if not specified. The port width is chosen to be the width of an equivalent rectangular waveguide as given in (9) of [1] and is computed as

$$W_{eff} = W - 1.08\frac{d^2}{p} + 0.1\frac{d^2}{W} \tag{2.35}$$

where d is the post diameter, p is the post period, and W is the center to center distance between the two post walls.

The first example is a rectangular substrate integrated waveguide with four metal posts inserted as shown in Fig. 2.8, where the two small posts have diameters of 0.22 mm, the two large posts have diameters of 0.77 mm, and those comprising the waveguide walls have diameters of 0.775mm. All the other dimensions and the material parameters are labeled in the figure. It acts as a bandpass filter that allows the electromagnetic waves between 27.42 to 28.75 GHz pass the device. Its S parameters are obtained and validated in Fig. 2.9. For manufacture convenience, the filter can also be implemented using metal posts of uniform diameters as shown in Fig. 2.10, where the two small metal posts are replaced by offset large ones, and all the metal posts have diameters of 0.775 mm. The S parameters are shown in Fig. 2.11. It is observed that the filters in Fig. 2.8 and Fig. 2.10 have the same characteristics. These two filters were firstly reported in [13], where a prototype of the filter in Fig. 2.10 was presented and is illustrated in Fig. 2.12. The second example is an equal

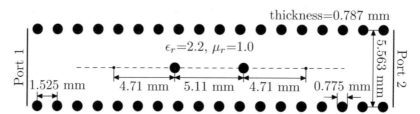

Figure 2.8: A bandpass filter with metal posts of different sizes.

power divider as shown in Fig. 2.13 with all the dimensions and material parameters labeled. The obtained results are verified in Fig. 2.14.

The third example as shown in Fig. 2.15 is a 3-dB coupler. Within the operating band from 22 to 26 GHz, most energy that injected into either Port 1 or Port 2 is equally split into Port 3 and Port 4. The simulated results are plotted in Fig. 2.16, where Fig. 2.16(a) demonstrates a good return loss for each input port as well as a good isolation between the two input ports, and Fig. 2.16(b) shows the equal power distribution for the two output ports.

More complicated circuit can be designed using this substrate integrated waveguide technique. For example, [11] uses two 3dB-couplers and two power dividers, which are similar to those in Fig. 2.13 and Fig. 2.15, respectively, as building blocks to design a six-port hybrid. A prototype is shown

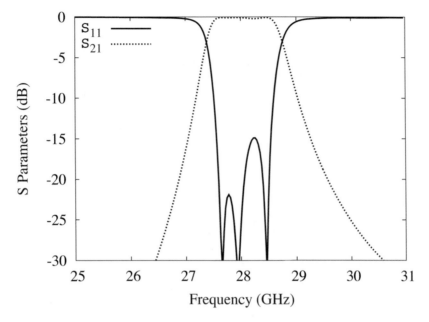

Figure 2.9: S parameters of the filter in Fig. 2.8.

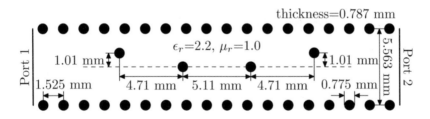

Figure 2.10: A bandpass filter with metal posts of the same sizes; from [10], copyright ©2008 IEEE.

in Fig. 2.17. When Port 1 or Port 2 is excited, the power is equally split into Port 3 to Port 6, but with different phase delay as shown in Table 2.1. This six-port hybrid is used in a direct receiver system for signaling [21]. Fig. 2.18 shows the prototype of a substrate integrated waveguide diplexer. When Port 1 is excited, the electromagnetic wave at the lower band is directed to Port 2 while the higher band goes to Port 3. The measured results are plotted in Fig. 2.19. Note that the two circuit examples in Fig. 2.17 and Fig. 2.18 can also be studied using the method presented in this chapter.

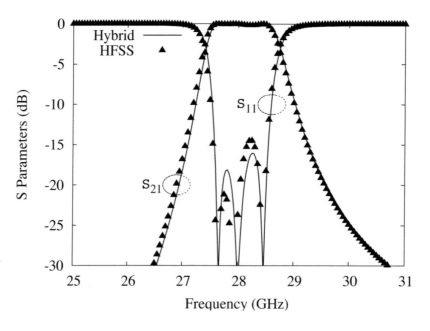

Figure 2.11: S parameters of the filter in Fig. 2.10; from [10], copyright ©2008 IEEE.

Table 2.1: Phase delay for a six-port hybrid; from [11].	
S parameter	**Phase delay**
S_{31}	$-180°$
S_{41}	$-90°$
S_{51}	$-90°$
S_{61}	$-180°$
S_{32}	$-135°$
S_{42}	$-135°$
S_{52}	$135°$
S_{62}	$45°$

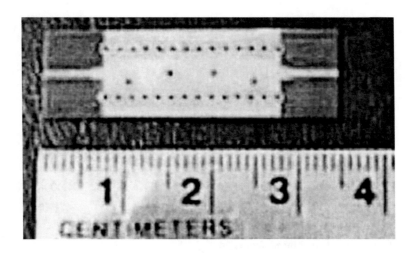

Figure 2.12: The prototype of the filter in Fig. 2.10; from [13], copyright ©2003 IEEE.

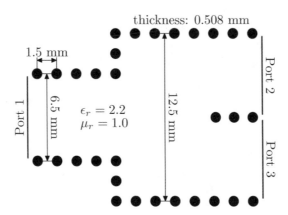

Figure 2.13: An equal power divider.

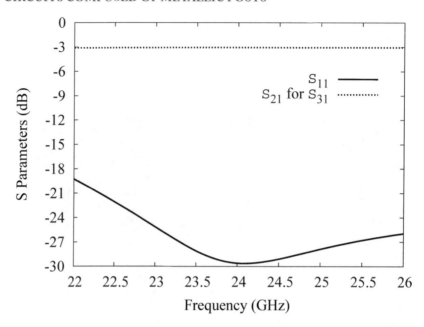

Figure 2.14: S parameters of the power divider in Fig. 2.13.

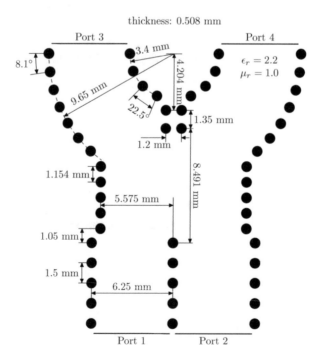

Figure 2.15: A 3-dB coupler.

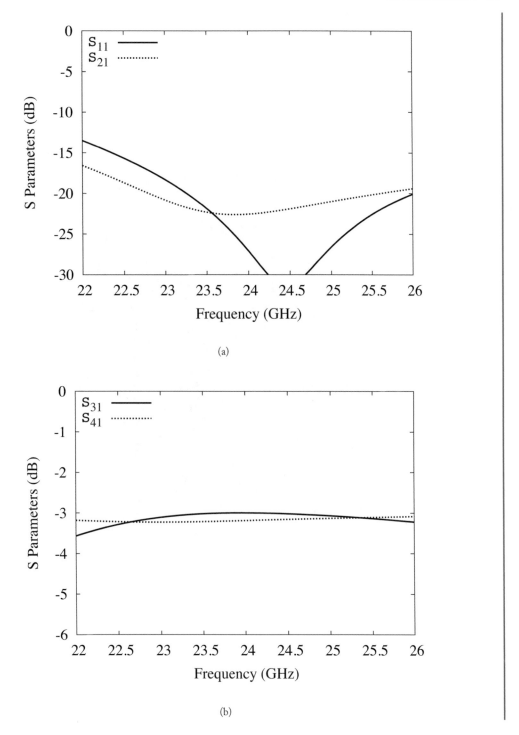

Figure 2.16: S parameters of the coupler in Fig. 2.15. (a) S_{11} and S_{21}, (b) S_{31} and S_{41}.

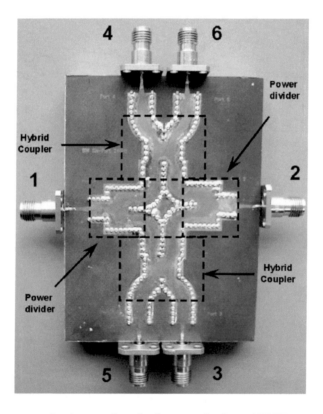

Figure 2.17: The prototype of a six port; from [11], copyright ©2005 IEEE.

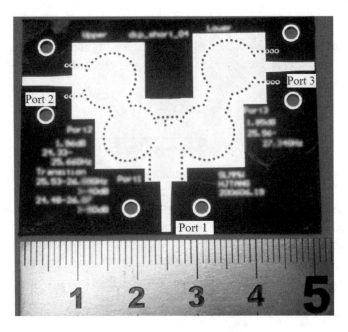

Figure 2.18: The prototype of a diplexer; from [12], copyright ©2007 IEEE.

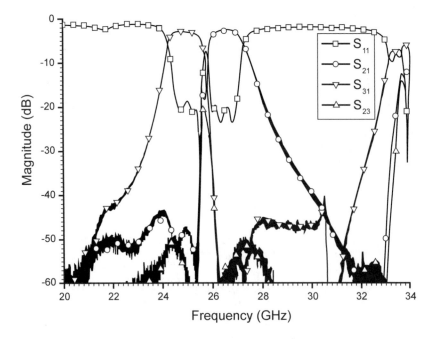

Figure 2.19: S parameters of the diplexer in Fig. 2.18; from [12], copyright ©2007 IEEE.

2.8   MODELING OF LOSSES

The loss due to the substrate and the metallic posts is readily modeled by setting nonzero surface impedance for the metallic posts, and taking substrate $\tan\delta$ into account in the calculation of permittivity and phase constant as

$$\epsilon = \epsilon_o\epsilon_r(1 - j\tan\delta) \tag{2.36}$$

$$\beta = 2\pi f\sqrt{\epsilon_o\mu_o\epsilon_r\mu_r(1 - j\tan\delta)} \tag{2.37}$$

where ϵ_o and μ_o are the permittivity and permeability of the free space. When the substrate is lossy, the computation of the Z matrix is the same as that described in Section 2.4, but that of the S matrix in Section 2.5 should be modified as the port impedance is a complex number but not a real number anymore. For ports with complex impedances, the voltages and currents can be written in terms of the scattering parameters as

$$V_{n\times 1} = \text{diag}\left(\frac{2R_1 - Z_1}{\sqrt{R_1}}, \cdots, \frac{2R_n - Z_n}{\sqrt{R_n}}\right)a_{n\times 1} + \text{diag}\left(\frac{Z_1}{\sqrt{R_1}}, \cdots, \frac{Z_n}{\sqrt{R_n}}\right)b_{n\times 1}$$

$$I_{n\times 1} = \text{diag}\left(\frac{1}{\sqrt{R_1}}, \cdots, \frac{1}{\sqrt{R_n}}\right)(a_{n\times 1} - b_{n\times 1}) \tag{2.38}$$

where Z_n is the impedance of the nth sub-port, and $R_n = \text{Re}(Z_n)$ is the real part of that impedance. Plugging (2.38) into

$$V_{n\times 1} = Z_{n\times n}I_{n\times 1} \tag{2.39}$$

gives

$$\text{diag}\left(\frac{2R_1 - Z_1}{\sqrt{R_1}}, \cdots, \frac{2R_n - Z_n}{\sqrt{R_n}}\right)a_{n\times 1} + \text{diag}\left(\frac{Z_1}{\sqrt{R_1}}, \cdots, \frac{Z_n}{\sqrt{R_n}}\right)b_{n\times 1}$$
$$= Z_{n\times n}\text{diag}\left(\frac{1}{\sqrt{R_1}}, \cdots, \frac{1}{\sqrt{R_n}}\right)(a_{n\times 1} - b_{n\times 1}). \tag{2.40}$$

Rearranging (2.40) results in

$$b_{n\times 1} = S_{n\times n}a_{n\times 1} \tag{2.41}$$

where

$$S_{n\times n} = \left[\text{diag}\left(\frac{Z_1}{\sqrt{R_1}}, \cdots, \frac{Z_n}{\sqrt{R_n}}\right) + Z_{n\times n}\text{diag}\left(\frac{1}{\sqrt{R_1}}, \cdots, \frac{1}{\sqrt{R_n}}\right)\right]^{-1}$$
$$\cdot\left[Z_{n\times n}\text{diag}\left(\frac{1}{\sqrt{R_1}}, \cdots, \frac{1}{\sqrt{R_n}}\right) - \text{diag}\left(\frac{2R_1 - Z_1}{\sqrt{R_1}}, \cdots, \frac{2R_n - Z_n}{\sqrt{R_n}}\right)\right]. \tag{2.42}$$

Note that (2.42) is reduces to (2.29) when $Z_n = R_n$.

As an example, if $\tan\delta = 0.003$ for the substrate and 0.5 Ω/square for the metallic posts are chosen, the S parameters of the circuit in Fig. 2.10 are validated in Fig. 2.20.

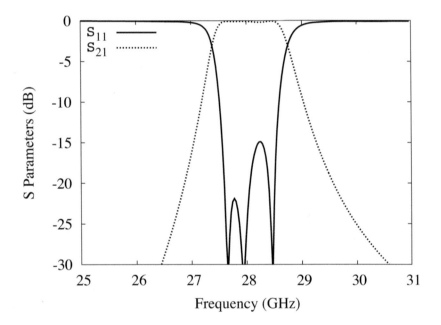

Figure 2.20: S parameters of the circuit in Fig. 2.10 with dielectric tanδ=0.003 and 0.5 Ω/square surface impedance for the metallic posts.

CHAPTER 3

SIW Circuits with Dielectric Posts

In Chapter 2, a method is presented to study a substrate integrated waveguide circuit with metallic posts inserted inside a homogeneous substrate. Although this type of circuit is the most widely adopted so far, a substrate integrated waveguide circuit with inhomogeneous substrate gives more freedom in the circuit design, and it may give some compact designs. This chapter generalizes the method presented in Chapter 2 to deal with a substrate integrated waveguide circuit with both dielectric and metallic cylinders. Herein, the field expression in each dielectric regions are required, and the boundaries conditions at all the material discontinuities including the boundaries of both metallic and dielectric cylinders should be enforced. The method is presented in a similar manner to that in Chapter 2, and two examples of inhomogeneous substrate integrated waveguide circuits are shown at the end of the chapter.

3.1 A TYPICAL SIW CIRCUIT AND ITS EQUIVALENT PROBLEM

Fig. 3.2 shows an x-y cut of a typical two-port substrate integrated waveguide circuit where all the black circular cylinders are metal posts and all the gray ones are made of dielectric materials. Specifically, the dielectric cylinder A and the metal post P are isolated from the other cylinders while the cylinder C and the metal post Q are embedded inside the cylinder B. The metal posts without labels emulate two metallic walls. Without losing generality, the circuit in Fig. 3.2 is used as an example to present the method.

Similar to Chapter 2, an approximated equivalent problem is constructed in two steps. First, a closed contour is generated to enclose the circuit as shown in Fig. 3.2, where part of the contour coincides the waveguide ports. Based on the surface equivalent principle, the electromagnetic fields inside the contour are due to the surface equivalent currents on the contour, with the presence of all the cylinders. Based on the fact that the field leakage from the post walls are very weak, the equivalent currents outside the post walls can be ignored. Therefore, the field inside the circuit is due to the equivalent currents at the ports, with the presence of all the cylinders as shown in Fig. 3.2(b).

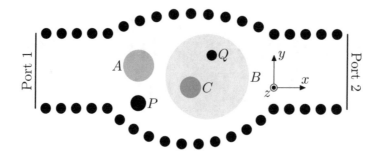

Figure 3.1: A typical substrate integrated waveguide circuit with dielectric posts.

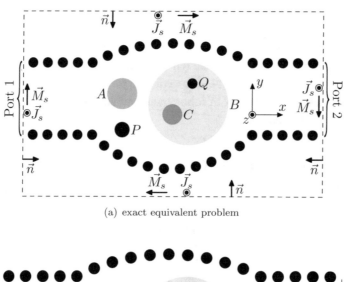

(a) exact equivalent problem

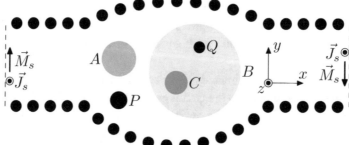

(b) approximated equivalent problem

Figure 3.2: Equivalent problems of a substrate integrated waveguide circuit with metallic and dielectric posts.

3.2   FIELD EXPRESSIONS IN DIFFERENT DIELECTRIC REGIONS

For the problem in Fig. 3.2(b), only the dielectric regions have non-zero field, and the field expressions are different for different regions. In cylinder A, the $\hat{z}$ directed electric field can be expanded using Bessel functions as

$$E_A^z = \sum_{m=-M_A}^{M_A} b_{m,A} J_m(\beta_A \rho_A) \exp(jm\phi_A) \tag{3.1}$$

where ρ_A and ϕ_A, referred to the local coordinates of the cylinder A as illustrated in Fig. 2.3, are the polar coordinates in a coordinate system whose origin is located at the center of cylinder A and whose x axis is parallel to the one shown in Fig. 3.2. β_A is the phase constant in cylinder A, and $b_{m,A}$ are unknown coefficients. Note that the expansion order m is truncated from $-M_A$ to M_A for practical computation. In cylinder C, the electric field is similar to (3.1), but replace all the subscripts A by C as

$$E_C^z = \sum_{m=-M_C}^{M_C} b_{m,C} J_m(\beta_C \rho_C) \exp(jm\phi_C). \tag{3.2}$$

In cylinder B, the electric field can be written as

$$E_B^z = \sum_{m=-M_B}^{M_B} b_{m,B} J_m(\beta_B \rho_B) \exp(jm\phi_B)$$

$$+ \sum_{\kappa=\{C,Q\}} \sum_{m=-M_\kappa}^{M_\kappa} a_{m,\kappa} H_m^{(2)}(\beta_B \rho_\kappa) \exp(jm\phi_\kappa) \tag{3.3}$$

where the field due to cylinder B is expanded using Bessel functions while those due to the cylinders C and Q are expanded using Hankel functions of the second kind. In (3.3), ρ_B and ϕ_B are the local coordinates of cylinder B, ρ_κ and ϕ_κ are those of cylinder κ where κ represents C or Q, and β_B is the phase constant in the medium of cylinder B. In the dielectric substrate, the region outside the cylinders A and B, the electric field can be expressed as

$$E_S^z = E_{\text{port}}^z + \sum_{n=1}^{N} \sum_{m=-M_n}^{M_n} a_{m,n} H_m^{(2)}(\beta_s \rho_n) \exp(jm\phi_n)$$

$$+ \sum_{\kappa=\{A,B,P\}} \sum_{m=-M_\kappa}^{M_\kappa} a_{m,\kappa} H_m^{(2)}(\beta_s \rho_\kappa) \exp(jm\phi_\kappa) \tag{3.4}$$

where E_{port}^z is the field due to the equivalent currents on the ports, N is the number of the metal posts that comprise the post walls, ρ_n and ϕ_n are the local coordinates of the nth metal post for the post walls, and β_s is the phase constant in the substrate.

E_{port}^z in (3.4) is derived in a MOM scheme in Chapter 2 following (2.2)–(2.6).

3.3   BOUNDARY CONDITIONS

For the problem in Fig. 3.2(b), three different types of boundary conditions are enforced. First, the tangential electric and magnetic fields at the boundary of a metal post are related using the surface impedance of the metal post as

$$\frac{E_S^z}{H_S^{\phi_P}} = R_P \text{ at } \rho_P = r_P \,, \quad \frac{E_B^z}{H_B^{\phi_Q}} = R_Q \text{ at } \rho_Q = r_Q$$

$$\frac{E_S^z}{H_S^{\phi_n}} = R_n \text{ at } \rho_n = r_n \text{ with } n = 1, \cdots, N \tag{3.5}$$

where $H_\alpha^{\phi_\kappa}$ with κ representing P or Q and α representing B or S is the ϕ component of the magnetic field in the local coordinate system of the cylinder κ and in the region represented by α. r_κ and R_κ are the radius and the surface impedance of the cylinder κ, respectively. $H_S^{\phi_n}$, r_n and R_n have the similar notations but are for the nth metal post on the post walls. Second, the tangential electric and magnetic fields are continuous on the boundary of a dielectric cylinder as

$$E_S^z = E_A^z \text{ and } H_S^{\phi_A} = H_A^{\phi_A} \text{ at } \rho_A = r_A$$

$$E_S^z = E_B^z \text{ and } H_S^{\phi_B} = H_B^{\phi_B} \text{ at } \rho_B = r_B$$

$$E_B^z = E_C^z \text{ and } H_B^{\phi_C} = H_C^{\phi_C} \text{ at } \rho_C = r_C \tag{3.6}$$

where $H_\alpha^{\phi_\kappa}$ and r_κ with κ representing A, B, or C and α representing A, B, C, or S have the similar notations as those in (3.5). Third, at the center of each sub-port, the $\hat{z}$ directed electric field equals to the equivalent magnetic current density as

$$E_S^z = I_M \tag{3.7}$$

if I_M is defined in the clockwise direction. Note that the first and the third boundary conditions in (3.5) and (3.7) are similar to those in (2.8) and (2.9), respectively. Again, there is no boundary conditions enforced on the enclosing box other than the ports.

Similar to a circuit with only metallic posts as discussed in Chapter 2, the equation (3.7) is ready to write into a linear combination of the unknown coefficients by substituting (3.4) into (3.7), but the equations in (3.5) and (3.6) are not. Although the electric field expressions required for (3.5) and (3.6) are given in (3.1)–(3.4), (3.3) and (3.4) are written as a combination of terms in different local coordinate systems and can not be used directly. Instead, they should be rewritten in the local coordinate system of the cylinder where the boundary conditions are applied. This can be done using the additional theorem of Bessel and Hankel functions. In Chapter 2, the additional theorem for a case where the observation cylinder is isolated from the source cylinder is given in (2.10). Herein, more cases are discussed considering the observation cylinder and and the source cylinder could

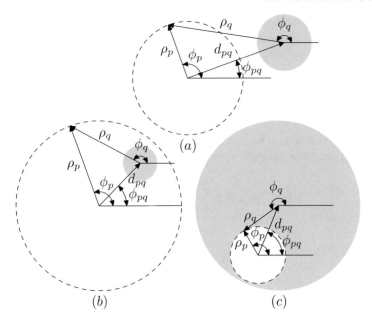

Figure 3.3: Different relative locations of the source and observation cylinders. (a) The source cylinder is isolated from the observation cylinder. (b) The source cylinder is enclosed inside the observation cylinder. (c) The observation cylinder is enclosed inside the source cylinder; from [10], copyright ©2008 IEEE.

be enclosed by each other. Fig. 3.3 shows three different relative locations of a gray cylinder and a cylinder in dashed line. The gray cylinder, referred to source cylinder q, scatters field, while the cylinder in dashed line, referred to observation cylinder p, is the one where the boundary conditions are applied. If the source cylinder is isolated from the observation cylinder as shown in Fig. 3.3(a) that is the same as Fig. 2.5, the electric field due to the source cylinder can be transformed into the local coordinate system of the observation cylinder using

$$H_n^{(2)}(\beta_s \rho_q) \exp(jn\phi_q) = \sum_{m=-Mp}^{Mp} J_m(\beta_s \rho_p) H_{m-n}^{(2)}(\beta_s d_{pq}) \exp(jm\phi_p)\Psi \qquad (3.8)$$

where $\Psi = \exp[j(n-m)\phi_{pq}]$. d_{pq} and ϕ_{pq} are the magnitude and phase of a vector pointing from the center of the observation cylinder p to that of the source cylinder p. ρ_p and ϕ_p are the local coordinates of the observation cylinder p, and ρ_q and ϕ_q are those of the source cylinder q. If the source cylinder is enclosed inside the observation cylinder as shown in Fig. 3.3(b), the electric field transformation can be conducted using

$$H_n^{(2)}(\beta_p \rho_q) \exp(jn\phi_q) = \sum_{m=-Mp}^{Mp} H_m^{(2)}(\beta_p \rho_p) J_{m-n}(\beta_p d_{pq}) \exp(jm\phi_p)\Psi \qquad (3.9)$$

where β_p is the phase constant in the observation cylinder p. If the observation cylinder is enclosed inside the source cylinder as shown in Fig. 3.3(c), the electric field transformation can be carried out using

$$J_n(\beta_q \rho_q) \exp(jn\phi_q) = \sum_{m=-M_p}^{M_p} J_m(\beta_q \rho_p) J_{m-n}(\beta_q d_{pq}) \exp(jm\phi_p) \Psi \tag{3.10}$$

where β_q is the phase constant in the source cylinder q. With (3.8)–(3.10), (3.3) and (3.4) can be rewritten in the local coordinate system of the observation cylinder p as

$$E_B^z(\rho_p, \phi_p) = \sum_{m=-M_p}^{M_p} V_m^p(\rho_p) \exp(jm\phi_p)$$

$$E_S^z(\rho_p, \phi_p) = \sum_{m=-M_p}^{M_p} U_m^p(\rho_p) \exp(jm\phi_p) \tag{3.11}$$

where $V_m^p(\rho_p)$ and $U_m^p(\rho_p)$ are linear combinations of the unknown coefficients and are independent of ϕ_p, and the super script p indicates that the expressions are dependent on the choice of the observation cylinder p. Specifically, in (3.4), the transformation of the electric field due to the ports follows (2.12)–(2.13). After the electric field is transformed, the ϕ component of the magnetic field in the local coordinate system of the cylinder p can be computed as

$$H_S^{\phi_p}(\rho_p, \phi_p) = \frac{-j}{\omega\mu} \frac{\partial E_S^z(\rho_p, \phi_p)}{\partial \rho_p} = \frac{-j}{\omega\mu} \sum_{m=-M_p}^{M_p} \exp(jm\phi_p) \frac{\partial U_m^p(\rho_p)}{\partial \rho_p} \tag{3.12}$$

for the substrate and

$$H_B^{\phi_p}(\rho_p, \phi_p) = \frac{-j}{\omega\mu} \frac{\partial E_B^z(\rho_p, \phi_p)}{\partial \rho_p} = \frac{-j}{\omega\mu} \sum_{m=-M_p}^{M_p} \exp(jm\phi_p) \frac{\partial V_m^p(\rho_p)}{\partial \rho_p} \tag{3.13}$$

for the region B. Both (3.12) and (3.13) are also series in terms of $\exp(jm\phi_p)$. Consequently, with the orthogonality of $\exp(jm\phi_p)$ for different m, a number of $2M_p + 1$ linear equations can be obtained from each equation in (3.5) and (3.6). For example, the first equation in (3.5) gives $2M_P + 1$ linear equations as

$$U_m^P(r_P) = \frac{-jR_P}{\omega\mu} \frac{\partial U_m^P(\rho_P)}{\partial \rho_P} \Big|_{\rho_P = r_P} \tag{3.14}$$

where $-M_P \leq m \leq M_P$. Each equation of the first row in (3.6) gives $2M_A + 1$ equations as

$$U_m^A(r_A) = b_{m,A} J_m(\beta_A r_A)$$

$$\frac{\partial U_m^A(\rho_A)}{\partial \rho_A} \Big|_{\rho_A = r_A} = b_{m,A} \beta_A J_m'(\beta_A r_A) \tag{3.15}$$

where $-M_A \leq m \leq M_A$ and $b_{m,A}$ is the expansion coefficient for the field in region A defined in (3.1). Each equation of the second row in (3.6) gives $2M_B + 1$ equations as

$$U_m^B(r_B) = V_m^B(r_B)$$
$$\frac{\partial U_m^B(\rho_B)}{\partial \rho_B} \Big|_{\rho_B = r_B} = \frac{\partial V_m^B(\rho_B)}{\partial \rho_B} \Big|_{\rho_B = r_B} \tag{3.16}$$

where $-M_B \leq m \leq M_B$.

3.4   *Z* MATRIX, *S* MATRIX AND SUB-PORT COMBINATION

Similar to what discussed in Chapter 2, for each metallic or dielectric cylinder, the number of equations obtained is equal to the number of unknowns associated with the that cylinder. But for a port, the number of equations is half of the number of unknowns. The computation of the Z matrix, S matrix and the sub-port combination are the same as those presented in Chapter 2. The derivations can be found in Sections 2.4-2.6.

3.5   EXAMPLES

The method is implemented to analyze a general substrate integrated waveguide circuit. The truncation of the expansion order for a cylinder κ has a similar expression to (2.34), and is given as

$$M_\kappa = \text{Int}(3\beta' a_\kappa + 1.5) \tag{3.17}$$

where the operator $\text{Int}(\cdot)$ gives the integer part of a real number, a_κ is the radius of the cylinder, and β' has different meanings for metallic and dielectric cylinders. For a metallic cylinder, β' is the real part of the phase constant in the medium where the cylinder is housed. For a dielectric cylinder, $\beta' = \max(\beta_{\text{in}}, \beta_{\text{out}})$ where $\max(\cdot, \cdot)$ gives the maximum value of two parameters, β_{in} is the real part of the phase constant inside the dielectric cylinder and β_{out} is that outside the cylinder. Several circuits are investigated using this method, and the results are verified by simulating each circuit with its physical thickness using HFSS. In the examples below, TE_{10} mode is assumed for all the ports, and zero surface impedance is assumed for each metallic post if not specified. The port width is chosen to be the width of an equivalent rectangular waveguide as given in (9) of [1].

The first example as shown in Fig. 3.4 is a bandpass filter with four circular dielectric cylinders embedded into a rectangular substrate integrated waveguide. The dielectric cylinders are made of a material with $\epsilon_r = 10.2$, and the dielectric substrate is with $\epsilon_r = 2.32$. The S parameters are shown in Fig. 3.5. The second example is a dual-band bandpass filter with four annular dielectric cylinders inserted into a rectangular substrate integrated waveguide as shown in Fig. 3.6. Each dielectric cylinder is made of a material with $\epsilon_r = 30.0$ and has a small hole at the center. The S parameters obtained by using the proposed method are plotted against those from HFSS in Fig. 3.7. It can be observed that for all the examples, the proposed method and HFSS give almost the same results.

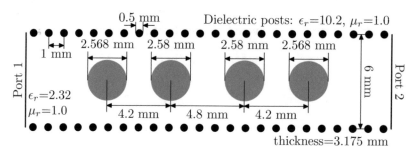

Figure 3.4: A bandpass filter with circular dielectric posts; from [10], copyright ©2008 IEEE.

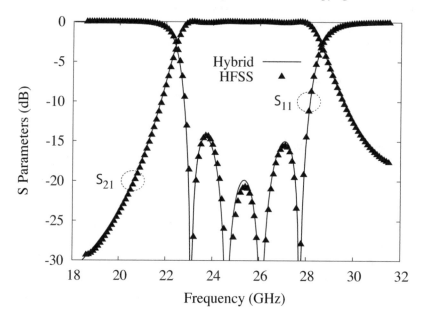

Figure 3.5: S parameters of the circuit in Fig. 3.4; [10], copyright ©2008 IEEE.

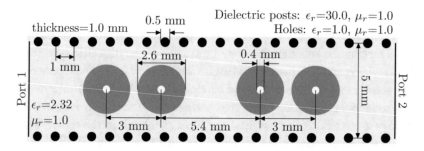

Figure 3.6: A dual-band bandpass filter with annular dielectric posts; [10], copyright ©2008 IEEE.

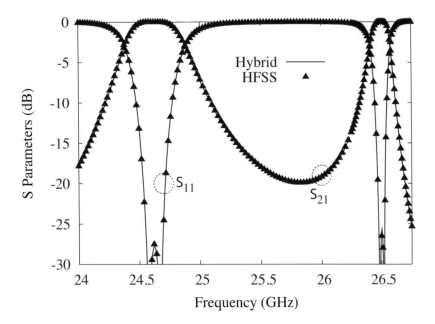

Figure 3.7: S parameters of the circuit in Fig. 3.6; from [10], copyright ©2008 IEEE.

CHAPTER 4

Even-Odd Mode Analysis of a Symmetrical Circuit

The method presented in Chapters 2 and 3 analyzes an arbitrary 2D substrate integrated waveguide circuit, where the unknown equivalent currents on all the geometrical discontinuities are considered in the formulation. For a geometrically symmetrical circuit, there are intrinsic relationships between the unknown currents that are symmetrically located. Those relationships can be used to reduce the number of unknowns when solving the problem, and, therefore, to speed up the simulation. One popular way of applying the symmetry is called even-odd mode analysis, by which the circuit is excited in specific ways such that the unknown currents that are symmetrically located are either equal or out-of-phase to each other. These excitations can be accomplished by inserting a PEC or PMC symmetry wall, and, therefore, it is sufficient to model only half of the circuit.

4.1 DECOMPOSITION OF A SYMMETRICAL CIRCUIT

Without losing generality, a four-port symmetrical circuit shown in Fig. 4.1 is used as an example to present the even-odd mode analysis. The broken line is the symmetry wall of the circuit. In order

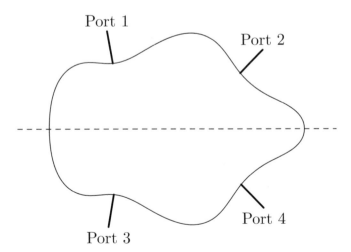

Figure 4.1: A typical four-port symmetrical circuit.

to get S_{11}, S_{21}, S_{31} and S_{41}, Port 1 is excited by incident voltage v as shown in Fig. 4.2. The S

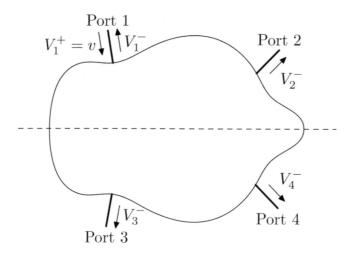

Figure 4.2: Excitation of Port 1 to get S parameters.

parameters can then be obtained by sensing the output voltage at all the four ports as

$$\begin{cases} S_{11} &= \dfrac{V_1^-}{v} \\[2mm] S_{21} &= \sqrt{\dfrac{R_1}{R_2}}\dfrac{V_2^-}{v} \\[2mm] S_{31} &= \sqrt{\dfrac{R_1}{R_3}}\dfrac{V_3^-}{v} \\[2mm] S_{41} &= \sqrt{\dfrac{R_1}{R_4}}\dfrac{V_4^-}{v} \end{cases} \tag{4.1}$$

where R_n is the impedance of the nth port, and V_n^- is the output voltage on that port assuming the port is connected to a matched load. The excitation of Port 1 can be decomposed into two sets of excitation as shown in Fig. 4.3. The excitation of the circuit in Fig. 4.3(a) is referred to as even mode excitation, and that in Fig. 4.3(b) is referred to as odd mode excitation. Considering linear devices only where the superposition principle holds, as the superposition of the excitations in Fig. 4.3(a) and Fig. 4.3(b) is the same as the excitation in Fig. 4.2, the superposition of the output in Fig. 4.3(a) and Fig. 4.3(b) should be the same as that in Fig. 4.2. Specifically,

$$V_n^- = V_n^{e-} + V_n^{o-} \tag{4.2}$$

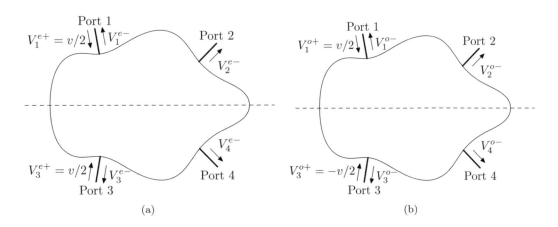

Figure 4.3: Decomposition of the excitation of Port 1 into (a) even mode excitation and (b) odd mode excitation.

where V_n^- is the output voltage at the nth port for the excitation shown in Fig. 4.2, V_n^{e-} is that at the nth port for the even mode excitation shown in Fig. 4.3(a), and V_n^{o-} is that at the same port for the odd mode excitation shown in Fig. 4.3(b). Plugging (4.2) into (4.1) gives

$$
\begin{cases}
S_{11} &= \dfrac{V_1^{e-} + V_1^{o-}}{v} \\[2ex]
S_{21} &= \sqrt{\dfrac{R_1}{R_2}} \dfrac{V_2^{e-} + V_2^{o-}}{v} \\[2ex]
S_{31} &= \sqrt{\dfrac{R_1}{R_3}} \dfrac{V_3^{e-} + V_3^{o-}}{v} \\[2ex]
S_{41} &= \sqrt{\dfrac{R_1}{R_4}} \dfrac{V_4^{e-} + V_4^{o-}}{v}.
\end{cases}
\tag{4.3}
$$

Due to the symmetry of the circuit and the in-phase excitation of Port 1 and Port 3 in Fig. 4.3(a), it is expected that $V_3^{e-} = V_1^{e-}$ and $V_4^{e-} = V_2^{e-}$. Similarly, due to the symmetry of the circuit and the out-of-phase excitation of Port 1 and Port 3 in Fig. 4.3(b), it is expected that $V_3^{o-} = -V_1^{o-}$ and $V_4^{o-} = -V_2^{o-}$. In addition, $R_1 = R_3$ and $R_2 = R_4$ because of the circuit symmetry. Therefore, (4.3) can be rewritten as

$$
\begin{cases}
S_{11} &= \dfrac{V_1^{e-} + V_1^{o-}}{v} \\[2ex]
S_{21} &= \sqrt{\dfrac{R_1}{R_2}}\,\dfrac{V_2^{e-} + V_2^{o-}}{v} \\[2ex]
S_{31} &= \dfrac{V_1^{e-} - V_1^{o-}}{v} \\[2ex]
S_{41} &= \sqrt{\dfrac{R_1}{R_2}}\,\dfrac{V_2^{e-} - V_2^{o-}}{v}.
\end{cases}
\tag{4.4}
$$

In the even mode excitation case in Fig. 4.3(a), the magnetic field on the symmetry wall is perpendicular to that wall due to the symmetry of the structure and the excitations. Therefore, only the up half of the circuit is needed to be analyzed by replacing the symmetry wall by a perfectly magnetic conductor (PMC) as shown in Fig. 4.4(a) as it keeps the same boundary conditions on the symmetry wall in Fig. 4.3(a). In the odd mode excitation case in Fig. 4.3(b), the magnetic field on the symmetry wall is parallel to that wall, and there is zero electric field on that wall due to the symmetry of the structure and the out-of-phase excitation. Thus, the symmetry wall can be replaced by a perfectly electric conductor (PEC) as shown in Fig. 4.4(b). Consequently, the analysis of Fig. 4.3(a) and Fig. 4.3(b) can be done by investigating Fig. 4.4(a) and Fig. 4.4(b), respectively, and (4.4) can then be rewritten as

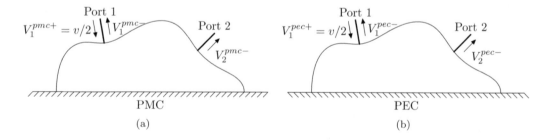

Figure 4.4: Analysis of (a) the even mode excitation case using PMC symmetry and (b) the odd mode excitation case using PEC symmetry.

$$
\begin{cases}
S_{11} &= \dfrac{V_1^{\text{PMC}-} + V_1^{\text{PEC}-}}{v} \\[2ex]
S_{21} &= \sqrt{\dfrac{R_1}{R_2}}\,\dfrac{V_2^{\text{PMC}-} + V_2^{\text{PEC}-}}{v} \\[2ex]
S_{31} &= \dfrac{V_1^{\text{PMC}-} - V_1^{\text{PEC}-}}{v} \\[2ex]
S_{41} &= \sqrt{\dfrac{R_1}{R_2}}\,\dfrac{V_2^{\text{PMC}-} - V_2^{\text{PEC}-}}{v}.
\end{cases}
\tag{4.5}
$$

where V_n^{PMC} is the output voltage at the nth port for the half circuit with PMC symmetry, and V_n^{PEC} is that for the half circuit with PEC symmetry, both assuming the ports are matched. The output voltage in Fig. 4.4 can be expressed as the product of the excitation and the S parameters as

$$\begin{cases} V_1^{\text{PMC}-} = v S_{11}^{\text{PMC}}/2 \\ V_2^{\text{PMC}-} = v S_{21}^{\text{PMC}}/2 \\ V_1^{\text{PEC}-} = v S_{11}^{\text{PEC}}/2 \\ V_2^{\text{PEC}-} = v S_{21}^{\text{PEC}}/2. \end{cases} \tag{4.6}$$

Substituting (4.6) into (4.5) leads to

$$\begin{cases} S_{11} = (S_{11}^{\text{PMC}} + S_{11}^{\text{PEC}})/2 \\ S_{21} = \sqrt{\dfrac{R_1}{R_2}}(S_{21}^{\text{PMC}} + S_{21}^{\text{PEC}})/2 \\ S_{31} = (S_{11}^{\text{PMC}} - S_{11}^{\text{PEC}})/2 \\ S_{41} = \sqrt{\dfrac{R_1}{R_2}}(S_{21}^{\text{PMC}} - S_{21}^{\text{PEC}})/2. \end{cases} \tag{4.7}$$

Equation (4.7) reveals how to obtain the S parameter of a symmetrical circuit from those of the two half circuits. For a $2N$-port symmetrical circuit as shown in Fig. 4.5 where the nth port and the n'th port with $1 \leq n \leq N$ are symmetrical to the broken line, the S parameters can be generalized as

$$\begin{cases} S_{mn} = S_{m'n'} = \sqrt{\dfrac{R_n}{R_m}}(S_{mn}^{\text{PMC}} + S_{mn}^{\text{PEC}})/2 \\ S_{mn'} = S_{m'n} = \sqrt{\dfrac{R_n}{R_m}}(S_{mn}^{\text{PMC}} - S_{mn}^{\text{PEC}})/2 \end{cases} \tag{4.8}$$

where the mth port and the nth port are located at the same side of the symmetry wall, and the n'th port is symmetrical to the nth port. Therefore, the analysis of a symmetrical circuit can be decomposed into the analyses of two half circuits. The method presented in Chapters 2 and 3 can be further generalized to deal with the symmetry walls.

4.2   HALF CIRCUIT WITH PMC SYMMETRY WALL

A typical half SIW circuit with PMC symmetry wall is shown in Fig. 4.6, where the circuit is composed of metallic posts, dielectric posts and a PMC wall. Similar to the problems shown in Fig. 2.1 and Fig. 3.1, an approximated equivalent problem is constructed so that the electromagnetic field inside the circuit is due to the the equivalent currents on the ports with the presence of all the cylinders. The field expressions and the boundary conditions are required to solve this problem.

The electric field in the cylinders A, B and C have the same forms as those for the problem in Fig. 3.1, and are given in (3.1), (3.3) and (3.2), respectively. Compared to (3.4), the electric field in the substrate for the problem in Fig. 4.6 has one more term due to the scattering from the PMC

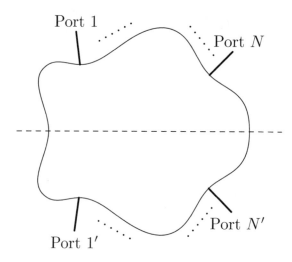

Figure 4.5: A typical $2N$-port symmetrical circuit.

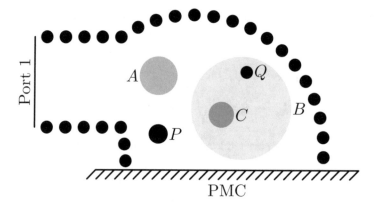

Figure 4.6: A typical half circuit with PMC symmetry.

wall, and is given as

$$E_S^z = E_{\text{port}}^z + E_{\text{PMC}}^z + \sum_{n=1}^{N} \sum_{m=-M_n}^{M_n} a_{m,n} H_m^{(2)}(\beta_s \rho_n) \exp(jm\phi_n)$$

$$+ \sum_{\kappa=\{A,B,P\}} \sum_{m=-M_\kappa}^{M_\kappa} a_{m,\kappa} H_m^{(2)}(\beta_s \rho_\kappa) \exp(jm\phi_\kappa) \qquad (4.9)$$

where E_{PMC}^z is the electric field due to the radiation from the equivalent magnetic current on the PMC wall, and can be computed in the same manner as that due to the magnetic current on a port using method of moment. The PMC wall is divided into several segments and each assumes a constant magnetic current. The electric field due to each segment has the same expression as (2.3). It can be evaluated numerically if the observation point is not located on the PMC segment itself, and is reduced to (2.6) if the observation point is exactly on the PMC segment. As will be seen later, there is no need to evaluate the electric field on the PMC wall when the boundary conditions are enforced. So, (2.6) is not required for the PMC wall.

The boundary conditions for the problem in Fig. 4.6 are similar to those for the problem in Fig. 3.2(b) except that one more condition should be enforced on the PMC wall. For convenience, all the boundary conditions are listed below as

$$\frac{E_S^z}{H_S^{\phi_P}} = R_P \text{ at } \rho_P = r_P , \quad \frac{E_B^z}{H_B^{\phi_Q}} = R_Q \text{ at } \rho_Q = r_Q$$

$$\frac{E_S^z}{H_S^{\phi_n}} = R_n \text{ at } \rho_n = r_n \text{ with } n = 1, \cdots , N \tag{4.10}$$

$$E_S^z = E_A^z \text{ and } H_S^{\phi_A} = H_A^{\phi_A} \text{ at } \rho_A = r_A$$
$$E_S^z = E_B^z \text{ and } H_S^{\phi_B} = H_B^{\phi_B} \text{ at } \rho_B = r_B$$
$$E_B^z = E_C^z \text{ and } H_B^{\phi_C} = H_C^{\phi_C} \text{ at } \rho_C = r_C \tag{4.11}$$

$$E_S^z = I_M \tag{4.12}$$

and

$$H_S^t = 0 \text{ on the PMC wall,} \tag{4.13}$$

where H_S^t is the magnetic field tangential to the PMC wall. Numerically, (4.13) is enforced on the center of each PMC segment. Note that the boundary conditions in (4.10)–(4.12) are the same as those for the problem in Fig. 3.2(b), and are also treated in the same manner as discussed in Chapter 3 using coordinate system transformation. It is because the magnetic current on the PMC wall in Fig. 4.6 can be treated in the same way as the magnetic current on the ports. But, the boundary condition (4.13) does need further investigation.

The H_t in (4.13) can be obtained by projecting the total magnetic field in the substrate onto the PMC wall. Similar to the electric field, the magnetic field in the substrate is due to the equivalent electric and magnetic currents on the port, the equivalent magnetic current on the PMC wall, and the scattering from cylinders A, B, P and the metal posts comprising the waveguide wall. The magnetic field can be derived from the electric field using

$$\vec{H} = \frac{j}{\omega\mu}\left(\frac{\hat{\rho}}{\rho}\frac{\partial E^z}{\partial \phi} - \hat{\phi}\frac{\partial E^z}{\partial \rho}\right). \tag{4.14}$$

Therefore, with the electric field in (4.9), the magnetic field in the substrate due to all the cylinders are

$$\vec{H}_{cylinder} = \frac{j}{\omega\mu_s}\left\{\sum_{n=1}^{N}\sum_{m=-M_n}^{M_n} a_{m,n}\exp(jm\phi_n)\left[\frac{\hat{\rho}_n}{\rho_n}jmH_m^{(2)}(\beta_s\rho_n) - \hat{\phi}_n\beta_s H_m^{(2)'}(\beta_s\rho_n)\right]\right.$$
$$\left. + \sum_{\kappa=\{A,B,P\}}\sum_{m=-M_\kappa}^{M_\kappa} a_{m,\kappa}\exp(jm\phi_\kappa)\left[\frac{\hat{\rho}_\kappa}{\rho_\kappa}jmH_m^{(2)}(\beta_s\rho_\kappa) - \hat{\phi}_\kappa\beta_s H_m^{(2)'}(\beta_s\rho_\kappa)\right]\right\} \quad (4.15)$$

where $\hat{\rho}_n$ and $\hat{\phi}_n$ are the two orthogonal unit vectors for the local coordinate system associated with the nth metal cylinder on the post wall, and $\hat{\rho}_\kappa$ and $\hat{\phi}_\kappa$ are those for the cylinder κ where κ represents A, B or P. The magnetic field due to an electric current as shown in Fig. 4.7 can be derived by put (2.2) in (4.14) as

$$\vec{H}_{J_s} = \frac{jI_J\beta_s}{4}\int_0^L \hat{\phi}' H_0^{(2)'}(\beta_s\rho')dx' \quad (4.16)$$

where $\hat{\phi}' = -\hat{x}\sin\phi' + \hat{y}\cos\phi'$. The derivatives of the Hankel function required for (4.15)

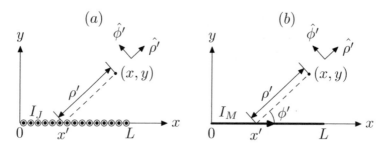

Figure 4.7: Radiation due (a) an electric current and (b) a magnetic current.

and (4.16) can be computed as

$$\frac{H_n^{(n)'}}{dx} = H_{n-1}^{(2)}(x) - \frac{n}{x}H_n^{(2)}(x). \quad (4.17)$$

For a magnetic current as shown in Fig. 4.7(b), the magnetic field can be derived by plugging (2.3) into (4.14) as

$$\vec{H}_{M_s}(x, y) = \frac{-\beta_s I_M}{4\omega\mu_s}\int_0^L\left[\frac{\hat{\rho}'}{\rho'}\cos\phi' H_1^{(2)}(\beta_s\rho') - \hat{\phi}'\sin\phi' H_1^{(2)'}(\beta_s\rho')\beta_s\right]dx'. \quad (4.18)$$

Using the derivative relation of the Hankel function in (4.17), $\vec{H}_{M_s}$ can be further derived as

$$\vec{H}_{M_s}(x, y) = \int_0^L I_M\vec{G}_{M_s}^H(x, y, x')dx' \quad (4.19)$$

where $G_{M_s}^H$ is the Green's function for the magnetic field due to a magnetic source and is written as

$$\vec{G}_{M_s}^H(x, y, x') = -\hat{\rho}'\sqrt{\frac{\epsilon_s}{\mu_s}}\frac{1}{4\rho'}\cos\phi' H_1^{(2)}(\beta_s\rho')$$

$$+ \hat{\phi}'\sin\phi'\left[\frac{\epsilon_s\omega}{4}H_0^{(2)}(\beta_s\rho') - \sqrt{\frac{\epsilon_s}{\mu_s}}\frac{1}{4\rho'}H_1^{(2)}(\beta_s\rho')\right] \qquad (4.20)$$

where $\rho' = \sqrt{(x-x')^2 + y'^2}$, $\hat{\rho}' = \hat{x}\cos\phi' + \hat{y}\sin\phi'$, $\hat{\phi}' = -\hat{x}\sin\phi' + \hat{y}\cos\phi'$, and ϵ_s as well as μ_s are the material parameters of the substrate. If the observation point (x, y) is not located on the magnetic current itself, (4.18) can be evaluated numerically. Otherwise, it should be derived separately as there is a singularity in the integrand when $\rho' = 0$. Suppose the observation point is exactly on the magnetic current, at an coordinate of $(x, 0)$ with $0 < x < L$, the magnetic field tangential to the magnetic current is derived in [22], and can be written as

$$H_{M_s}^t(x, 0) = I_M\frac{\epsilon_s^2}{\mu_s^2}\frac{E_{J_s}^z(x, 0)}{I_J} + \frac{1}{4}\sqrt{\frac{\epsilon_s}{\mu_s}}\left\{H_1^{(2)}(\beta_s x) + H_1^{(2)}[\beta_s(L-x)]\right\} \qquad (4.21)$$

where $E_{J_s}^z(x, 0)/I_J$ is the electric field due a unit electric current and can be computed following (2.4)– (2.5). Therefore, by using (4.15)–(4.21), H^t in (4.13) can be written as a polynomial in all the unknowns. A set of linear equations can be derived from (4.10)– (4.13), and the circuit characteristics can be obtained following the same procedure presented in Sections 2.4-2.6.

4.3   HALF CIRCUIT WITH PEC SYMMETRY WALL

A typical half SIW circuit with PEC symmetry wall is similar to that shown in Fig. 4.6 but it is replacing the PMC wall by a PEC wall as shown in Fig. 4.8. The electric field expressions have the

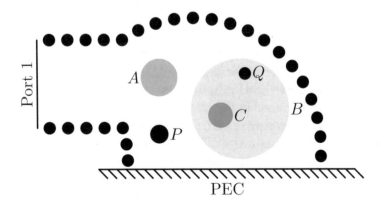

Figure 4.8: A typical half circuit with PEC symmetry.

same form as that for Fig. 4.6 except that in the substrate, the field due to the PMC wall is replaced by the field due to the PEC wall as

$$E_S^z = E_{port}^z + E_{PEC}^z + \sum_{n=1}^{N} \sum_{m=-M_n}^{M_n} a_{m,n} H_m^{(2)}(\beta_s \rho_n) \exp(jm\phi_n)$$

$$+ \sum_{\kappa=\{A,B,P\}} \sum_{m=-M_\kappa}^{M_\kappa} a_{m,\kappa} H_m^{(2)}(\beta_s \rho_\kappa) \exp(jm\phi_\kappa). \tag{4.22}$$

The boundary conditions in (4.10)–(4.12) remain for the problem in Fig. 4.8, but (4.13) should be replaced by

$$E_S^z = 0 \text{ on the PEC wall,} \tag{4.23}$$

which will be evaluated at the center of each PEC wall segment. The boundary conditions in (4.10)–(4.12) can be expanded in the same way as described in Chapter 2 because the additional electric current on the PEC wall can be treated in the same way as that on the port. The boundary condition (4.23) can be written in a polynomial in all the unknowns in the similar way to (4.12) by setting $I_M = 0$ in (4.12). Similar to the analysis of the half circuit with PMC wall, the circuit characteristics can be derived following Sections 2.4-2.6.

4.4   EXAMPLES

Fig. 4.9 shows an SIW 180° hybrid, also called a ring hybrid. All the dimensions and material parameters are labeled in the figure. When electromagnetic power is injected into Port 1, half of the power goes to Port 2 and another half goes to Port 3. The output at Port 2 and Port 3 are in phase. When electromagnetic power is injected into Port 2, it is equally split and directed to Port 1 and Port 4, but with opposite phases. As it is a symmetrical circuit, it can be investigated by simulating two half circuits as discussed before. Fig. 4.10 illustrates the half circuit of Fig. 4.9 with the PMC symmetry wall and Fig. 4.11 shows the resulting S parameters. Fig. 4.12 illustrates the half circuit with the PEC symmetry wall, and the resulting S parameters are plotted in Fig. 4.13. The S parameters of the entire circuit in Fig. 4.9 can then be obtained by combining the results in Fig. 4.11 and Fig. 4.13 using (4.8). The final results are verified by the full wave analysis if the entire circuit using HFSS, as shown in Fig. 4.14 for magnitude and in Fig. 4.15 for phase.

If the symmetry wall passes across a port, that port should be split into two symmetrical ports before applying the even-odd mode analysis. Fig. 4.16 shows a symmetrical T junction designed using the substrate integrated waveguide technique, where the symmetry wall goes through Port A. In order to apply the even-odd mode analysis, Port A is split into two symmetrical ports Port 1 and Port 1′ as shown in Fig. 4.17. The field distributions of Port 1 and Port 1′ are kept the same as if they were not split. Now, two half circuits as shown in Fig. 4.18 are constructed and analyzed separately. Based on (4.8), the S parameters of the circuit in Fig. 4.17 can then be obtained by combining the

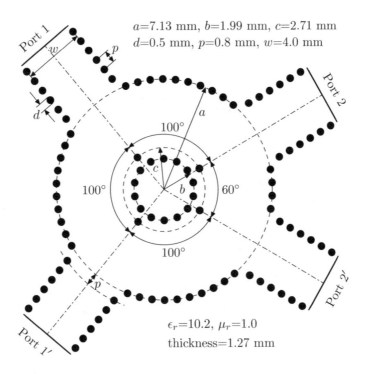

a=7.13 mm, b=1.99 mm, c=2.71 mm
d=0.5 mm, p=0.8 mm, w=4.0 mm

ϵ_r=10.2, μ_r=1.0
thickness=1.27 mm

Figure 4.9: An SIW ring hybrid; from [10], copyright ©2008 IEEE.

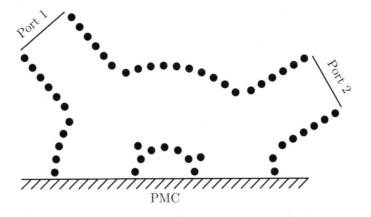

PMC

Figure 4.10: Analysis of the even mode excitation of the ring hybrid in Fig. 4.9 using PMC symmetry.

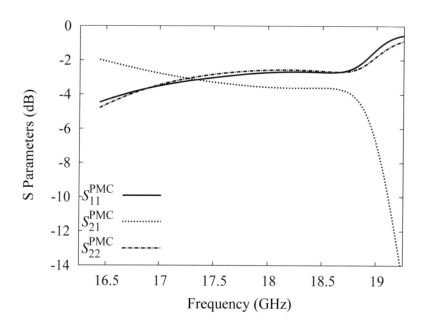

Figure 4.11: S parameters of the half circuit with a PMC wall shown in Fig. 4.10.

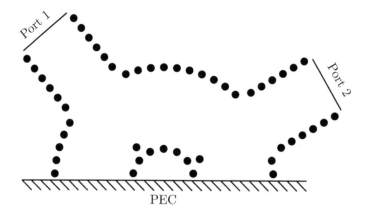

Figure 4.12: Analysis of the odd mode excitation of the ring hybrid in Fig. 4.9 using PEC symmetry.

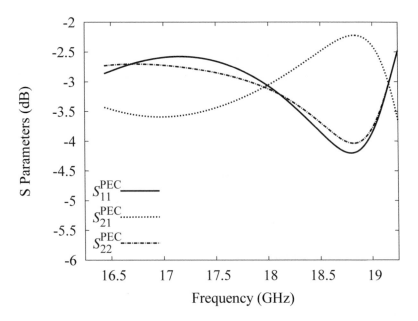

Figure 4.13: S parameters of the half circuit with a PEC wall shown in Fig. 4.12.

results for Fig. 4.18(a) and Fig. 4.18(b). Subsequently, the final S parameters for the circuit in 4.16 can be obtained in a similar manner to the sub-port combination in (2.31). Specifically,

$$|S_{AA}| = \sqrt{\frac{|S_{11} + S_{11'}|^2 + |S_{1'1} + S_{1'1'}|^2}{2}}$$

(4.24)

$$|S_{AB}| = \sqrt{|S_{12}|^2 + |S_{1'2}|^2}.$$

(4.25)

Both the simulated results obtained by using and not using the circuit symmetry are plotted and compared in Fig. 4.19, where good agreement between this two methods is observed.

The even-odd mode analysis breaks a large symmetrical circuit into two smaller circuits, and then probably speeds up the simulation. Given a symmetrical circuit with $2N$ unknowns, each of its half circuits has $N + N_s$ unknowns where N_s is the number of unknowns on the symmetry wall. The efficiency of the even-odd mode analysis depends on the topology of the circuit. If the circuit is much longer in the direction perpendicular to the symmetry wall than in the direction along the wall, the even-odd mode analysis will speed up the simulation a lot as it tremendously reduce the number of unknowns. Otherwise, the even-odd mode analysis may require longer time and is not a good choice. For the ring hybrid in Fig. 4.9, the analysis of the entire circuit takes 275 seconds to

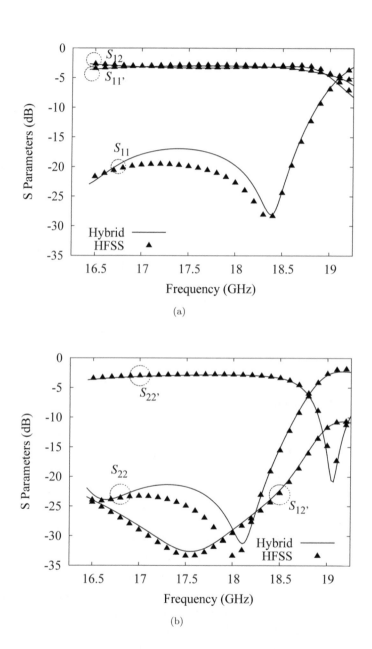

Figure 4.14: S parameters of the circuit in Fig. 4.9. (a) S_{11}, S_{12} and $S_{11'}$, (b) S_{22}, $S_{22'}$ and $S_{12'}$; from [10], copyright ©2008 IEEE.

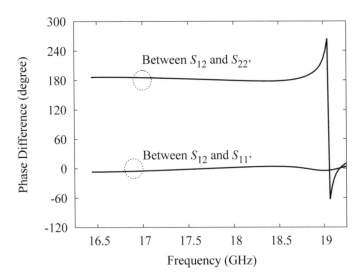

Figure 4.15: Phase difference between the ports for the circuit in Fig. 4.9.

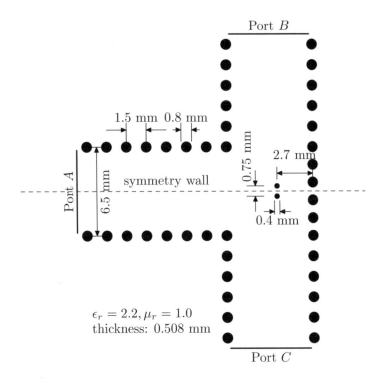

Figure 4.16: A substrate integrated waveguide T junction.

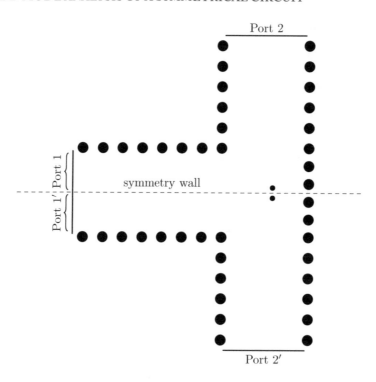

Figure 4.17: Split Port *A* of Fig. 4.16 into two symmetrical ports.

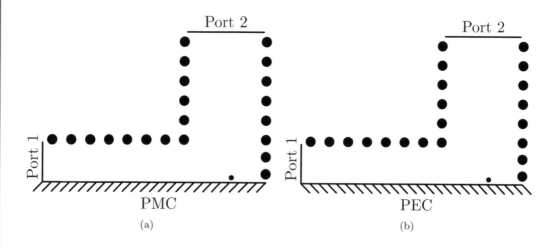

Figure 4.18: Analyses of the (a) even mode excitation using PMC wall and (b) odd mode excitation using PEC wall for the T junction in Fig. 4.17.

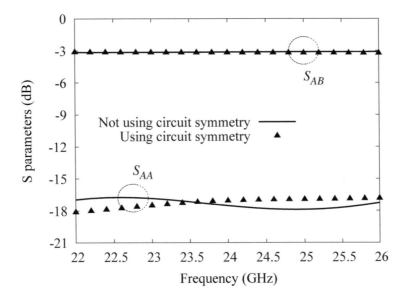

Figure 4.19: S parameters of the T junction in Fig. 4.16.

simulate 20 frequency points from 16.4 to 19.1 GHz, and the even-odd mode analysis only takes 176 seconds on the same computer. For the T junction in Fig. 4.16, the even-odd mode analysis takes 70 seconds to simulate 20 frequency points from 22 to 26 GHz, while the simulation of the entire structure takes the same amount of time of 71 seconds. The improvement in efficiency is not as significant as the ring hybrid case because the T junction does not extend that much in the direction perpendicular to the symmetry wall.

CHAPTER 5

Microstrip to SIW Transition and Half Mode SIW

In the previous chapters, a substrate integrated waveguide circuit is excited through waveguide ports. It is a good choice to get the circuit characteristics if the circuit is connected to another waveguide circuit. However, in many situations, a conventional microstrip transmission line interface is required. For example, in S parameters measurements, the substrate integrated waveguide is often connected to an end mounted SMA connector though an SIW-to-microstrip transition. Another situation is that when active components such as a transistor is incorporated into an SIW circuit, microstrip line is also required to connect the transistor. There are several types of SIW to microstrip transition proposed, such as a tapered line and a DC decoupled transition [15]. The tapered line is the most straightforward way to connect a substrate integrated waveguide and a microstrip. It can be approximated as a 2D problem by modeling the boundary of the transition as a PMC wall, and, therefore, it can be analyzed by the method presented in this book. This chapter shows how to solve a substrate integrated waveguide circuit with tapered line transitions. Furthermore, the method presented in this chapter can be applied to half mode substrate integrated waveguide, which is the compact counterpart of the substrate integrated waveguide. Several examples of this type of half mode circuits are shown at the end of this chapter.

5.1   METHOD

Fig. 5.1 illustrates a typical two-port substrate integrated waveguide circuit with each port connected to a tapered line transition. All the cylinders have the same meaning as those in Fig. 3.1, and the solid line is the boundary of the copper sheet on the top of the substrate. The ground plane is located on the bottom side of the substrate. Strictly speaking, the problem in Fig. 5.1 is a three-dimensional problem as the field around the edge of the copper sheet varies in the z direction, and it cannot be solved by a 2D code as presented in this book. But, if the substrate is electrically thin, which is normally true, it is still a good approximation to treat it as a 2D problem.

On the bottom side of the copper sheet, the electric current flows along the edge at the boundary of the copper sheet, and thus the current component normal to the edge equals to zero. With the relationship of the current and the magnetic field as

$$\vec{J}s = \vec{n} \times \vec{H},$$

(5.1)

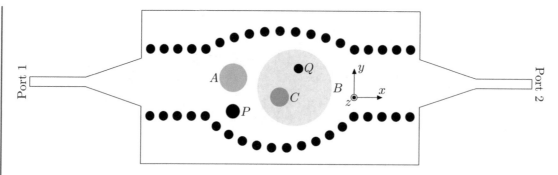

Figure 5.1: A typical SIW circuit with SIW-to-microstrip transition.

we have

$$J_p = H_t = 0 \qquad (5.2)$$

at the sheet boundary, where J_p is the current density on the copper sheet and normal to the sheet boundary, and H_t is the magnetic field along the sheet boundary as illustrated in Fig. 5.2. Therefore,

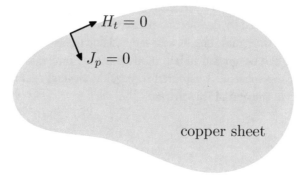

Figure 5.2: The current and magnetic field at the boundary of a copper sheet.

the boundary of the copper sheet is equivalent to a PMC wall with the same contour as they share the same boundary condition. Consequently, the problem in Fig. 5.1 is equivalent to the problem in Fig. 5.3. Furthermore, the equivalent magnetic current on the PMC wall outside the SIW wall is weak and can be ignored. So, Fig. 5.3 is reduced to Fig. 5.4. As a substrate integrated waveguide circuit with PMC walls, the circuit in Fig. 5.4 is similar to that in Fig. 4.6, and can be solved using the method presented in Section 4.2. Note that the problem in Fig. 5.3 can also be solved, but Fig. 5.4 requires less number of unknowns and is sufficient to give an accurate result.

For a microstrip port, the port impedance required for computing S parameters as discussed in Section 2.5 is different to that of the rectangular waveguide port as their field distributions across

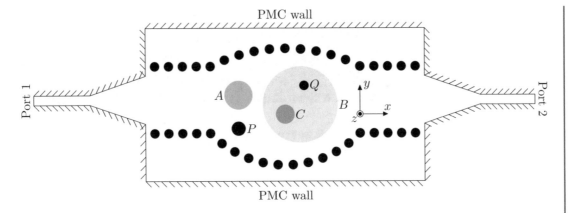

Figure 5.3: The equivalent problem of the circuit in Fig. 5.1.

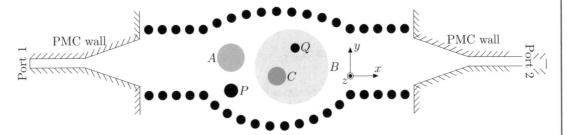

Figure 5.4: The approximated equivalent problem of the circuit in Fig. 5.1.

the port are different. If there is only one sub-port for each physical microstrip port, the impedance equals to the characteristic impedance of the microstrip line. Otherwise, the impedance of the nth sub-port should be computed as

$$R_n = R_o \frac{L}{l_n} \tag{5.3}$$

where R_o is the characteristic impedance of the microstrip port, L is the width of the physical port, and l_n is the width of the nth sub-port.

5.2   TRANSITION EXAMPLES

Fig. 5.5 shows the top view of a substrate integrated waveguide filter. It is similar to that shown in Fig. 2.10 but with SIW-to-microstrip transitions. An equivalent problem is constructed as shown in Fig. 5.6 where the boundary of the circuit is replaced by PMC walls. It can be analyzed using the method presented before. Considering the fringing field at the edge of the copper sheet, in Fig. 5.6, the PMC wall is expanded from the physical boundary with a distance of h, where h is the thickness

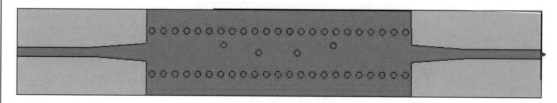

Figure 5.5: A substrate integrated waveguide filter with SIW-to-microstrip transitions.

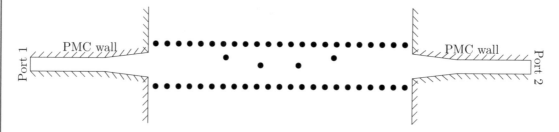

Figure 5.6: Simulation model of the filter in Fig. 5.5.

of the substrate. For the port in Fig. 5.6, it assumes that the power flow is constant across the port. Therefore, if a port is divided into several sub-ports, the power distribution required for (2.31) can be written as

$$p_i = \frac{l_i}{L} \tag{5.4}$$

where l_i is the length of the ith sub-port and L is the total length of the physical port. The simulated results for the filter in Fig. 5.5 are plotted in Fig. 5.7 and they are verified by simulating the exact structure using HFSS. Good agreement can be observed.

The SIW-to-microstrip transition can also be used to excite the high order mode for a substrate integrated waveguide. Fig. 5.8 shows a coplanar microstrip transition, which was firstly reported in [23]. When the two microstrip ports are excited in-phase and with the same power level, the TE_{20} mode will be excited in the substrate integrated waveguide as shown in Fig. 5.9. An ultra-high data rate can be achieved by transmitting two independent data streams simultaneously, one over the TE_{10} mode and the another over the TE_{20} mode.

Another SIW-to-microstrip transition is shown in Fig. 5.10. Since the DC component is decoupled between the input and output ports, it can be directly used to connect an active circuit that requires a DC bias. The field distribution is illustrated in Fig. 5.11, which shows that the electromagnetic energy is coupled through the inter-digital structure. It should be pointed out that this transition cannot be simulated using the 2D method discussed in this book because the electromagnetic field around the inter-digital structure is not uniform in the direction normal to the substrate anymore.

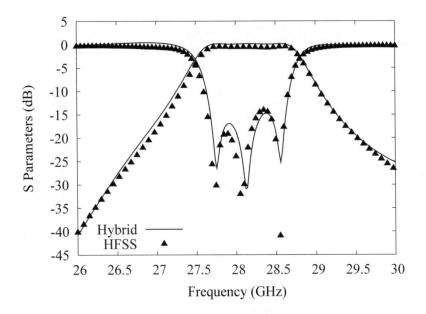

Figure 5.7: S parameters of the filter in Fig. 5.5.

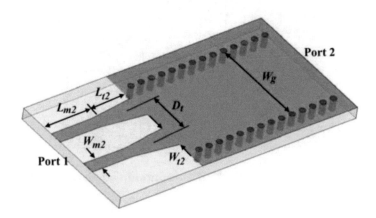

Figure 5.8: A coplanar microstrip TE_{20} transition; from [23], copyright ©2009 IEEE.

5.3   HALF MODE SIW

In last section, it shows that it is a good approximation to model the boundary of a copper sheet as a PMC wall. In other words, this give a convenient way to realize the PMC wall by cutting the copper sheet. This idea can be applied to halve a circuit size while keeping its electrical characteristics. Chapter 4 demonstrates that the even-mode excitation of a symmetrical circuit is equivalent to a

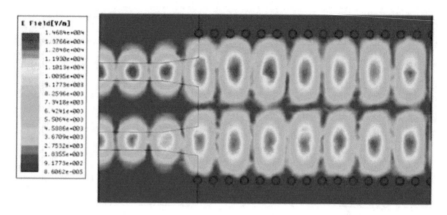

Figure 5.9: The electric field distribution of the circuit in Fig. 5.8; from [23], copyright ©2009 IEEE.

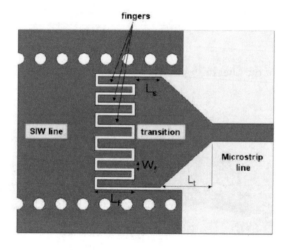

Figure 5.10: A DC-decoupled transition between microstrip to substrate integrated waveguide; from [15], copyright ©2007 IEEE.

half structure with PMC symmetry wall. A half structure can be implemented by cutting the copper sheet at a position where the PMC wall is located. Fig. 5.12 illustrates such a so called half mode substrate integrated waveguide and its equivalent full size counterpart. It can be observed that the field distribution of the half mode circuit is similar to that of full circuit. It implies that they will demonstrate similar network characteristics. The half mode waveguide concept can be applied to design compact circuit. Fig. 5.13 shows a 180° 3-dB coupler, where each branch of the circuit is a half mode substrate integrated waveguide. Its size is, therefore, the half of its full size counterpart. When Port 1 or Port 4 is excited, the input power is equally directed to Port 2 and Port 3, but with

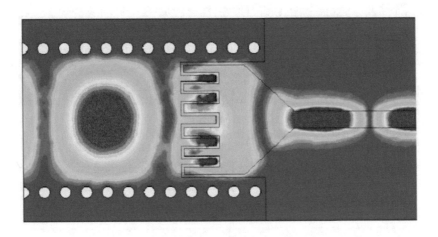

Figure 5.11: Field distribution of the transition in Fig. 5.10; from [15], copyright ©2007 IEEE.

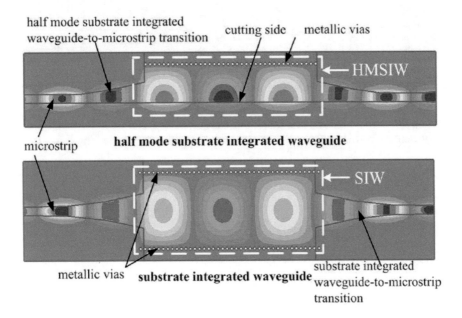

Figure 5.12: A half mode substrate integrated waveguide circuit and its equivalent full substrate integrated waveguide circuit; from [24], copyright ©2007 IEEE.

a 180° phase difference. The phase difference is achieved by the via fins as shown in Fig. 5.13. The

Figure 5.13: A half mode 180° 3-dB coupler; from [24], copyright ©2007 IEEE.

field distribution of the coupler is illustrated in Fig. 5.14 for the excitation of Port 1. The 180° phase difference can be observed from the field distribution as the field peak at Port 2 is half wavelength shifted from that of the Port 3. More half mode circuits are show in Fig. 5.15.

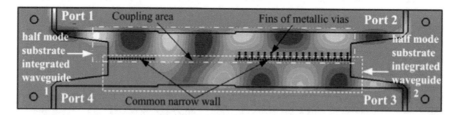

Figure 5.14: Field distribution of the half mode coupler in Fig. 5.13; from [24], copyright ©2007 IEEE.

Note that the circuits in Fig. 5.8, Fig. 5.12, Fig. 5.13 and Fig. 5.15 can also be analyzed using the method presented in this book, in the same manner as for the filter in Fig. 5.5.

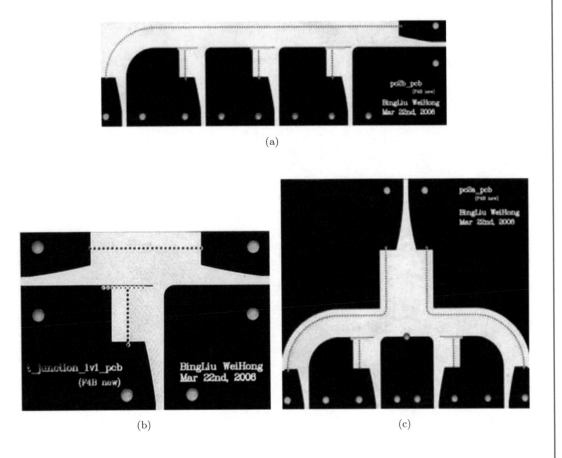

Figure 5.15: More half mode substrate integrated waveguide circuits, (a) a series power divider, (b) a T junction and (c) a four-way power divider; from [25], copyright ©IEEE.

CHAPTER 6

SIW Antennas

So far, most substrate integrated waveguide circuits discussed in this book can be analyzed using the presented 2D method. But, the substrate integrated waveguide technique itself is not limited to 2D designs only. Specifically, radiation elements such as a slot carved on the copper sheet or an open aperture at the edge of the substrate can be incorporated into the design to make an antenna. Electromagnetic energy will radiate through those openings in a controlled manner. For these kinds of structure, a three-dimensional algorithm is required and the 2D method introduced in this book is not applicable anymore. However, for many substrate integrated waveguide antenna arrays, the 2D method is still valuable to efficiently design the feeding network. In this chapter, several antenna examples based on the substrate integrated waveguide technique are demonstrated.

6.1   SIW HORN ANTENNAS

Because a substrate integrated waveguide emulates a conventional rectangular waveguide on a printed circuit board, some waveguide based antennas can also be implemented using the substrate integrated waveguide technique. Considering an H-plane sectoral horn is formed by flaring a waveguide in the plane normal to the electrical field, a similar design can be realized using the substrate integrated waveguide technique.

Fig. 6.1 shows two H-plane sectoral horn designed based on the substrate integrated waveguide technique. They are loaded by dielectric substrate of either rectangular shape or elliptical shape to match the free space. A four-element linear array using this type of radiating element is shown in Fig. 6.2, and gives a radiation pattern at 27 GHz as plotted in Fig. 6.3.

A monopulse antenna with eight radiating elements is shown in Fig. 6.4. The feeding network is designed such that the two ports excites the radiating elements in different ways. For one port, referred to as a sum port, all the horns are excited in phase so it gives a maximum radiation in the broadside direction of the array. While for the other port, referred to as a difference port, half of the horns are excited out of phase to the other half, so it gives a null radiation in the broadside direction. Fig. 6.5 and Fig. 6.6 shows the radiation patterns for the excitation of the sum port and difference port, respectively. This antenna can be used in an monopulse tracking system.

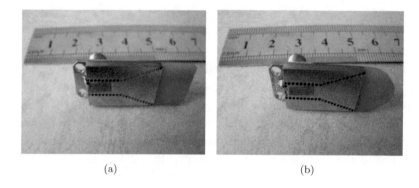

(a)   (b)

Figure 6.1: Substrate integrated waveguide H-plane horn loaded by (a) rectangular dielectric material and (b) elliptical dielectric material; from [17], copyright ©2010 IEEE.

Figure 6.2: A four-element substrate integrated waveguide H-plane horn array; from [17], copyright ©2010 IEEE.

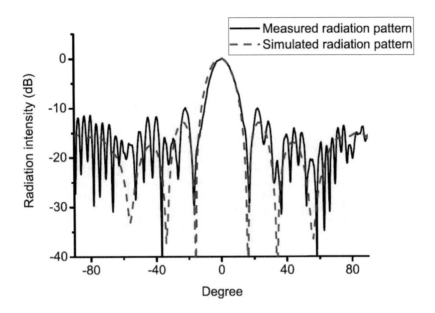

Figure 6.3: Radiation pattern of the array in Fig. 6.2 at 27 GHz; from [17], copyright ©2010 IEEE.

Figure 6.4: A substrate integrated waveguide monopulse antenna; from [17], copyright ©2010 IEEE.

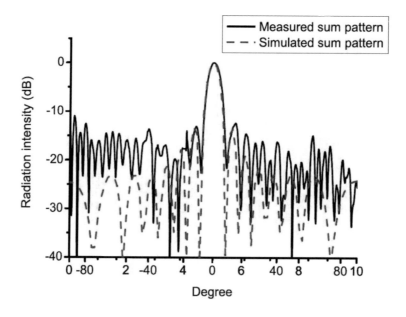

Figure 6.5: Sum pattern of the monopulse antenna in Fig. 6.4; from [17], copyright ©2010 IEEE.

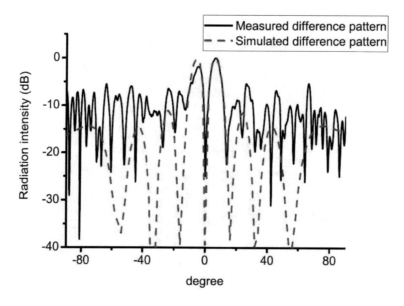

Figure 6.6: Difference pattern of the monopulse antenna in Fig. 6.4; from [17], copyright ©2010 IEEE.

6.2   SIW SLOT ANTENNAS

Inspired by the traditional waveguide slot antenna, it is straightforward to introduce radiating slots on the copper sheet of a substrate integrated waveguide to make an antenna. Fig. 6.7 shows the front and back views of an omnidirectional antenna operating at 6 GHz, where radiating slots are cut on both sides of the substrates. Its radiation pattern in both E and H planes are plotted in Fig. 6.8.

Figure 6.7: An omnidirectional antenna; from [16], copyright ©2008 IEEE.

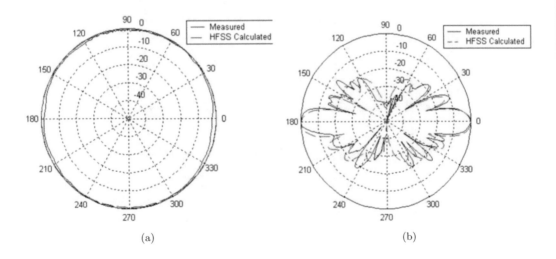

Figure 6.8: Radiation patterns of the antenna in Fig. 6.7 at 6 GHz for (a) E plane and (b) H plane; from [16], copyright ©2008 IEEE.

Fig. 6.9 shows a circularly polarized traveling wave slot antenna operating at 16 GHz. The circular polarized radiation is achieved by introducing offset slots in orthogonal directions. The axial ratio in the broadside direction of the array is plotted in Fig. 6.10, and the radiation pattern is shown in Fig. 6.11.

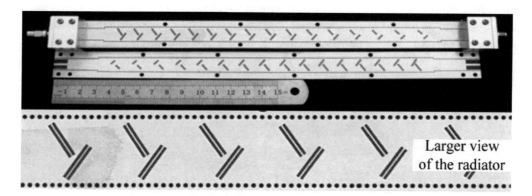

Figure 6.9: An circularly polarized antenna array; from [26], copyright ©2009 IEEE.

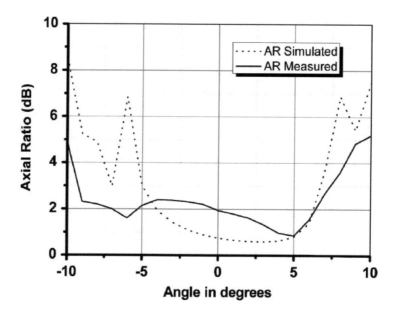

Figure 6.10: Axial ratio of the antenna in Fig. 6.9, in the broadside direction of the array; from [26], copyright ©2009 IEEE.

In Section 5.3, half mode substrate integrated waveguide is realized by cutting the copper sheet along the center of the waveguide to emulate a PMC symmetry wall. The same concept can

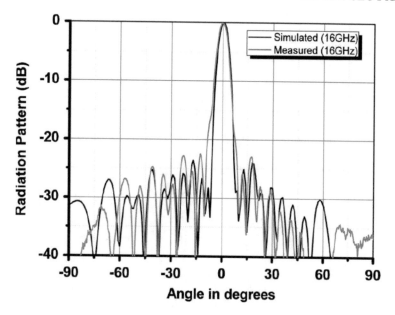

Figure 6.11: Radiation pattern of the antenna in Fig. 6.9; from [26], copyright ©2009 IEEE.

be used to design a half mode slot antenna. Fig. 6.12 shows such an antenna at Ka-band. The return loss and radiation pattern are plotted in Fig. 6.13 and Fig. 6.14, respectively.

Figure 6.12: A halfmode slot antenna array; from [27], copyright ©2009 IEEE.

More complicated antennas can be implemented using the substrate integrated waveguide technique. Fig. 6.15 shows a leaky-wave antenna fed by a dual reflector system, where the reflectors

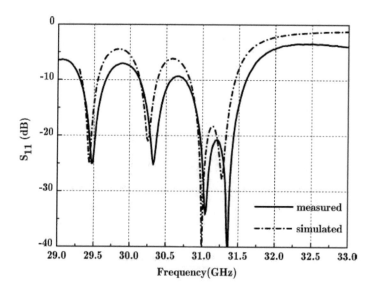

Figure 6.13: S parameters of the antenna in Fig. 6.12; from [27], copyright ©2009 IEEE.

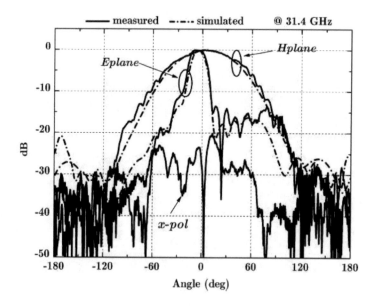

Figure 6.14: Radiation patterns of the antenna in Fig. 6.12; from [27], copyright ©2009 IEEE.

are implemented by metal posts. The feeding reflectors and the radiating slots are fabricated in a single circuit board. The reflection coefficient and radiation patterns are shown in Fig. 6.16.

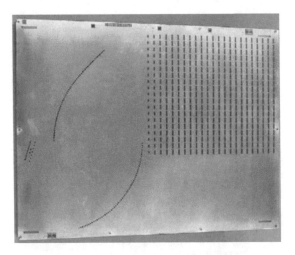

Figure 6.15: A dual reflector leaky wave antenna; from [28], copyright ©2008 IEEE.

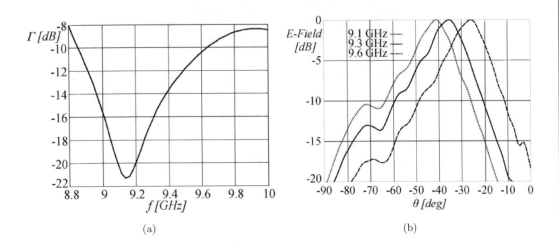

Figure 6.16: Measured results of the antenna in Fig. 6.15 for (a) reflection coefficient and (b) radiation patterns; from [28], copyright ©2008 IEEE.

Fig. 6.17 shows a slot antenna array that is fed by a rotman lens, where the radiation part is composed of nine longitudinal slot arrays. This antenna is used for digital beam scanning. The excitation of a port from B1 to B7 introduces different phase delays for each slot array, thus tilting the radiation beam in the E-plane. Fig. 6.18 illustrates the pattern scanning of the antenna for exciting port B1 to B7. In addition, an array of four such antennas arranged in a way as shown in Fig. 6.19 is able to scan in both the E and H planes.

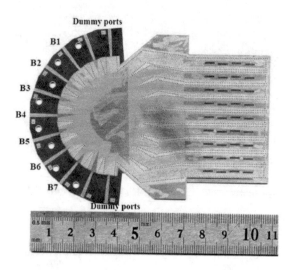

Figure 6.17: A substrate integrated waveguide rotman lens and its slot antenna array; from [18], copyright ©2008 IEEE.

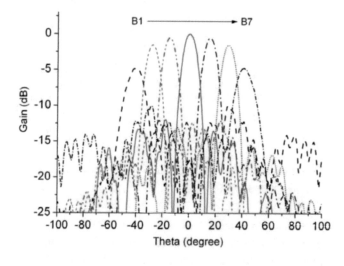

Figure 6.18: Radiation patterns of the antenna in Fig. 6.17; from [18], copyright ©2008 IEEE.

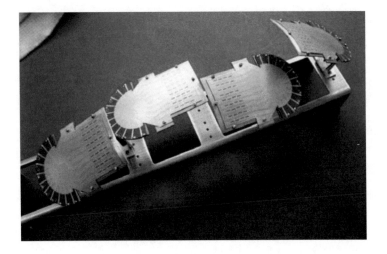

Figure 6.19: An array of four rotman-lens fed slot antenna to scan both the E and H planes; from [18], copyright ©2008 IEEE.

Bibliography

[1] F. Xu and K. Wu, "Guided-wave and leakage characteristics of substrate integrated waveguide", *IEEE Trans. Microwave Theory Tech.*, vol. 53, pp. 66-72, Jan. 2005. DOI: 10.1109/TMTT.2004.839303 1, 5, 16, 33

[2] J. Hirokawa and M. Ando, "Single-layer feed waveguide consisting of posts for plane TEM wave excitation in parallel plates", *IEEE Trans. Antennas Propagat.*, vol. 46, pp. 625-630, May 1998. DOI: 10.1109/8.668903 1

[3] D. Deslandes and K. Wu, "Accurate modeling, wave mechanisms, and design considerations of a substrate integrated waveguide", *IEEE Trans. Microwave Theory Tech.*, vol. 54, pp. 2516-2526, June 2006. DOI: 10.1109/TMTT.2006.875807 1, 5

[4] T. Okoshi, "Planar Circuits for Microwaves and Lightwaves," Springer-Verlag, 1985. 2

[5] D. Pissoort, E. Michielssen, F. Olyslager and D. De Zutter, "Fast Analysis of 2-D Electromagnetic Crystal Devices Using a Periodic Green Function Approach", *J. Lightw. Technol.*, vol. 23, no. 7, pp. 2294-2308, Jul. 2005. DOI: 10.1109/JLT.2005.850029 2

[6] D. Pissoort, E. Michielssen and A. Grbic, "An Electromagnetic Crystal Green Function Multiple Scattering Technique for Arbitrary Polarizations, Lattices, and Defects," *J. Lightw. Technol.*, vol. 25, no. 2, pp. 571-583, Feb. 2007. DOI: 10.1109/JLT.2006.889364 2

[7] S. Boscolo and M. Midrio, "Three-Dimensional Multiple-Scattering Technique for the Analysis of Photonic-Crystal Slabs", *J. Lightw. Technol.*, vol. 22, no. 12, pp. 2778-2786, Dec. 2004. DOI: 10.1109/JLT.2004.833276 2

[8] D. Pissoort, E. Michielssen, D. Vande Ginste and F. Olyslager, "Fast-Multipole Analysis of Electromagnetic Scattering by Photonic Crystal Slabs," *J. Lightw. Technol.*, vol. 25, no. 9, pp. 2847-2863, Sep. 2007. DOI: 10.1109/JLT.2007.902771 2

[9] X. H. Wu and A. A. Kishk, "A Hybrid Method to Study the Substrate Integrated Waveguide Circuit", Asia-Pacific Microwave Conference, Bangkok, Thailand, 11-14 December 2007, pp. 541-544. DOI: 10.1109/APMC.2007.4554970 2

[10] X. H. Wu and A. A. Kishk, "Hybrid of Method of Moments and Cylingdrical Eigenfunction Expansion to Study Substrate Integrated Waveguide Circuits", *IEEE Trans. Microwave Theory and Techniques*, vol. 56, pp. 2270-2276, August, 2008. DOI: 10.1109/TMTT.2008.2004255 2, 8, 15, 17, 18, 31, 34, 35, 47, 50

[11] X. Xu, R. G. Bosisio, and K. Wu, "A new six-port junction based on substrate integrated waveguide technology", *IEEE Trans. Microwave Theory Tech.*, vol. 53, pp. 2267-2273, July 2005. DOI: 10.1109/TMTT.2005.850455 2, 16, 18, 22

[12] H. J. Tang, W. Hong, J.-X. Chen, G. Q. Luo and K. Wu, "Development of millimeter-wave planar diplexers based on complementary characters of dual-mode substrate integrated waveguide filters with circular and elliptic cavities", *IEEE Trans. Microwave Theory Tech.*, vol. 55, pp. 776-782, April 2007. DOI: 10.1109/TMTT.2007.893655 2, 23

[13] D. Deslandes and K. Wu, "Single-substrate integration technique of planar circuits and waveguide filters," *IEEE Trans. Microwave Theory Tech.*, vol. 51, pp. 593-596, Feb. 2003. DOI: 10.1109/TMTT.2002.807820 2, 16, 19

[14] W. Che, K. Deng, E. K. N. Yung, and K. Wu, "H-plane 3-dB hybrid ring of high isolation in substrate-integrated rectangular waveguide (SIRW)," *Microw. Opt. Technol. Lett.*, vol. 48, no. 3, pp. 502-505, 2006. DOI: 10.1002/mop.21392 2

[15] M. Abdolhamidi, A. Enayati, M. Shahabadi and R. Faraji-Dana, "Wideband Single-Layer DC-Decoupled Substrate Integrated Waveguide (SIW)-to-Microstrip Transition Using an Interdigital Configuration", *Proceedings of Asia-Pasific Microwave Conference*, Dec. 11-14 2007, pp. 1-4. DOI: 10.1109/APMC.2007.4555065 2, 55, 60, 61

[16] G. Hua, W. Hong, X. H. Sun and H. X. Zhou, "Design of An Omnidirectional Line Array with SIW Longitudinal Slot Antenna", *International Conference on Microwave and Millimeter Wave Technology*, 2008, pp. 1114-1117. DOI: 10.1109/ICMMT.2008.4540620 2, 69

[17] H. Wang, D. G. Fang, B. Zhang and W. Q. Che, "Dielectric Loaded Substrate Integrated Waveguide (SIW) H-Plane Horn Antennas", *IEEE Trans. Antennas and Propagation*, vol. 58, no. 3, Mar. 2010, pp. 640–647. 2, 66, 67, 68

[18] J. J. Cheng, W. Hong, K. Wu, Z. Q. Kuai, C. Yu, J. X. Chen, J. Y. Zhou and H. J. Tang, "Substrate Integrated Waveguide (SIW) Rotman Lens and Its Ka-Band Multibeam Array Antenna Applications", *IEEE Trans. Antennas and Propagation*, vol. 56, no. 8, Aug. 2008, pp. 2504-2513. DOI: 10.1109/TAP.2008.927567 2, 74, 75

[19] "HFSS, 10.0." Ansoft Corporate, Pittsburgh, PA. 2

[20] A. A. Kishk and P.-S. Kildal, "Asymptotic Boundary Conditions for Strip-loaded Scatterers Applied to Circular Dielectric Cylinders Under Oblique Incidence," *IEEE Trans. Antennas Propagat.*, vol. 45, pp. 51-56, Jan. 1997. DOI: 10.1109/8.554240 5

[21] X. Xu, R. G. Bosisio and K. Wu, "Analysis and Implementation of Six-Port Software-Defined Radio Receiver Platform", *IEEE Trans. Microwave Theory and Techniques*, vol. 54, no. 7, pp. 2937-2943, July, 2006. DOI: 10.1109/TMTT.2006.877449 17

[22] Paul Mallette Goggans, "A Combined Method of Moments and Approximate Boundary Condition Solution for Scattering from a Conducting Body with a Dielectric Filled Cavity", Phd dissertation, Auburn University, 1990. 45

[23] A. Suntives and R. Abhari, "Design and Application of Multimode Substrate Integrated Waveguide in Parallel Multichannel Signaling System", *IEEE Trans. Microwave Theory and Techniques*, vol. 57, no. 6, pp. 1563-1571, Jun. 2009. DOI: 10.1109/TMTT.2009.2020777 58, 59, 60

[24] B. Liu, W. Hong, Y. Zhang, H. J. Tang, X. Yin and K. Wu, "Half Mode Substrate Integrated Waveguide 180° 3-dB Directional Couplers", *IEEE Trans. Microwave Theory and Techniques*, vol. 55, no. 12, pp. 2586-2592, December, 2007. DOI: 10.1109/TMTT.2007.909749 61, 62

[25] B. Liu, W. Hong, L. Tian, H.-B. Zhu, W. Jiang and K. Wu, "Half Mode Substrate Integrated Waveguide (HMSIW) Multi-way Power Divider", *Proceedings of Asia-Pasific Microwave Conference*, Dec. 12-15 2006, pp. 917-920. DOI: 10.1109/APMC.2006.4429562 63

[26] P. Chen, W. Hong, Z. Kuai and J. Xu, "A substrate Integrated Waveguide Circular Polarized Slot Radiator and Its Linear Array", *IEEE Antennas and Wireless Propagation Letters*, vol. 8, pp. 120-123, 2009. DOI: 10.1109/LAWP.2008.2011062 70, 71

[27] Q. H. Lai, W. Hong, Z. Q. Kuai, Y. S. Zhang and K. Wu, "Half-Mode Substrate Integrated Waveguide Transverse Slot Array Antennas", *IEEE Transactions on Antennas and Propagation*, vol. 57, no. 4, April, 2009, pp. 1064-1072. DOI: 10.1109/TAP.2009.2015799 71, 72

[28] M. Ettorre, A. Neto, G. Gerini and S. Maci, "Leaky-Wave Slot Array Antenna Fed by a Dual Reflector System", *IEEE Transactions on Antennas and Propagation*, vol. 56, no. 10, Oct. 2008, pp. 3143-3149. DOI: 10.1109/TAP.2008.929457 73

Authors' Biographies

XUAN HUI WU

Xuan Hui Wu received the B.Eng. degree from the Department of Information Science & Electronic Engineering (ISEE), Zhejiang University, China, in 2001, the M.Eng. degree from the Department of Electrical and Computer Engineering (ECE), National University of Singapore, Singapore, in 2005, and the Ph.D. degree from the Department of Electrical Engineering, University of Mississippi, USA.

From 2002 to 2004, he is with the Institute for Infocomm Research (I^2R), Agency of Science Technology and Research, Singapore, as a Research Graduate Student. From 2004 to 2009, he is with the Microwave Research Laboratory, University of Mississippi. Since 2009, he is with Radio Waves Inc. , Billerica, Massachusetts. His current research interests include computational electromagnetics, optimizations in electromagnetics, ultrawide-band radio systems and multiple-input-multiple-output systems. Dr. Wu is a member of Sigma Xi Society and a member of Phi Kappa Phi Society.

AHMED A. KISHK

Ahmed A. Kishk is a Professor of Electrical Engineering, University of Mississippi (since 1995). He was an Associate Editor of Antennas & Propagation Magazine from 1990 to 1993. He is now an Editor of Antennas & Propagation Magazine. He was a co-editor of the special issue on Advances in the Application of the Method of Moments to Electromagnetic Scattering Problems in the ACES Journal. He was also an editor of the ACES Journal during 1997. He was an Editor-in-Chief of the ACES Journal from 1998 to 2001. He was the chair of Physics and Engineering division of the Mississippi Academy of Science (2001-2002). He was a guest editor of the special issue on artificial magnetic conductors, soft/hard surfaces, and other complex surfaces, on the IEEE Transactions on Antennas and Propagation, January 2005.

His research interest includes the areas of design of millimeter frequency antennas, feeds for parabolic reflectors, dielectric resonator antennas, microstrip antennas, soft and hard surfaces, phased array antennas, and computer aided design for antennas. He has published over 150 refereed Journal articles and book chapters. He is a coauthor of the Microwave Horns and Feeds book (London, UK, IEE, 1994; New York: IEEE, 1994) and a coauthor of Chapter 2 on Handbook of Microstrip Antennas (Peter Peregrinus Limited, United Kingdom, edited by J. R. James and P. S. Hall, Ch. 2, 1989). Dr. Kishk received the 1995 and 2006 outstanding paper award for papers published in the Applied Computational Electromagnetic Society Journal. He received the 1997

Outstanding Engineering Educator Award from Memphis section of the IEEE. He received the Outstanding Engineering Faculty Member of the 1998. He received the Award of Distinguished Technical Communication for the entry of IEEE Antennas and Propagation Magazine, 2001. He received the 2001 and 2005 Faculty research award for outstanding performance in research. He also received The Valued Contribution Award for outstanding Invited Presentation, "EM Modeling of Surfaces with STOP or GO Characteristics - Artificial Magnetic Conductors and Soft and Hard Surfaces" from the Applied Computational Electromagnetic Society. He received the Microwave Theory and Techniques Society Microwave Prize 2004. Dr. Kishk is a Fellow member of IEEE since 1998 (Antennas and Propagation Society and Microwave Theory and Techniques), a member of Sigma Xi society, a member of the U.S. National Committee of International Union of Radio Science (URSI) Commission B, a member of the Applied Computational Electromagnetics Society, a member of the Electromagnetic Academy, and a member of Phi Kappa Phi Society.